HANDHELD COMPUTERS FOR CHEFS

HANDHELD COMPUTERS FOR CHEFS

SUSAN SYKES HENDEE

MOHAMMAD AL-UBAYDLI

WILEY

John Wiley & Sons, Inc.

For general information on our other products and services or for technical support,
please contact our Customer Care Department within the United States at (800) 762-
2974, outside the United States at (317) 572-3993 or fax (317) 572-4002.

Wiley also publishes its books in a variety of electronic formats. Some content that
appears in print may not be available in electronic books. For more information about
Wiley products, visit our web site at www.wiley.com.

Library of Congress Cataloging-in-Publication Data:
Hendee, Susan Sykes, 1951–
    Handheld computers for chefs / Susan Sykes Hendee, Mohammad Al-Ubaydli.
       p.   cm.
    Includes index.
    ISBN: 978-0-471-78919-2 (pbk.)
1. Restaurant management—Data processing.   2. Pocket computers.   3. Cooks—
Information services   I. Al-Ubaydli, Mohammad.   II. Title.
    TX911.3.E4H46 2008
    647.95068—dc22   2006037548

Printed in the United States of America

10 9 8 7 6 5 4 3 2 1

Contents

This part explains that handhelds are not futuristic gadgets, just to-day's technology working well for chefs around the world. The following chapters take the user through all of the possible applications of such machines.

Preface

From: Susan Hendee
To: Mohammad Al-Ubaydli
Subject: Re Handhelds

I was wondering if you would ever consider working with someone on a paperback *Handheld Computers for Chefs*, building on your success *Handheld Computers for Doctors*?

I am intrigued by the table of contents from your book.

Susan Sykes Hendee Ph.D., CCE
Associate Professor

Neither author knew it in 2004, but chefs have a lot in common with doctors. Both groups are overworked professionals responsible for teams of workers with a range of skills performing a variety of tasks. Both have to deal with huge amounts of information and inventory, and that means dealing with stress and lots of it.

A great chef handles stress with grace and poise to deliver great meals for the patrons. But like doctors around the world, chefs are beginning to use handheld computers to reduce that stress and improve the service for their customers.

The goal of this book is to teach you how to make such improvements in your kitchen.

Every day, Susan opens the diary on her handheld computer. Unlike her friends' paper versions, which are littered with tiny scribbles and repeated crossing outs, the entry for Susan's day always appears clear. Changing appointments is easy, and scheduling them as regular events takes no effort. As she goes through the day, she will pick up the names of good books and recipes for her classes, and add them to her to-do list.

She has entire courses of lecture notes in her pocket. She wrote them during lectures she attended, or on the beach while planning her own lectures. Her notes are often made faster and always more structured than those of her "papered" colleagues. Visits to clients' sites are also a place for culinary pearls of wisdom, which she scribbles down eagerly. For heavy levels of text input, she often switches from handwriting to typing—she carries a pocket-sized keyboard, which unfolds into a full-sized keyboard.

When she sees fellow students with a handheld computer, she beams those lecture notes to them. Beaming involves lining up the two devices and seamlessly transferring information using infrared. You really have to try this to appreciate its elegance.

When Susan works at food preparation institutions, she can document yields, assemble shopping lists, and track inventory using her handheld. As she visits institutions around the United States, she sees food preparation teams where each member uses their handheld computer to record and share such data.

Despite all of this efficiency, food preparation involves a lot of waiting. Susan no longer finds this irritating. She buys books that she never thought she would have time to read and devours them during such waiting times. She downloads the latest newspapers, which she digests at her own pace.

Then there are the games.

About the Book

This book is for anyone trying to cope with the information overload in the culinary industry. This includes chefs, restaurant owners, culinary teachers, and culinary students.

From this book, we hope you will realize that handheld technology can make a real difference in your work today, so you will be encouraged to buy a device. We also want to give you the tools to help your colleagues get a handheld, and take advantage of their group-work features.

To achieve such changes in your working life, you do not have to be good with computers, or even like them. We recommend handhelds to everyone because we know that anyone can learn to use them.

How to Use This Book

We have tried to keep this book short and entertaining. This is because we want you to enjoy reading it, and then enjoy setting up your own handheld projects. The book is divided into three sections:

Part 1 "Why *Star Trek* Is Science Past": Many people around the world think that *Star Trek* is good science fiction. However, it was only after several million Palm Pilots were sold that we saw episodes in the series featuring actors writing on small handheld computers. *Star Trek* is not really about the future, just about today's technology working a little faster.[1] We want you to think of handhelds in the same way. They are not futuristic gadgets, just today's technology working well for busy people including chefs around the world.

Part 2 "Why Two Handhelds Are Better Than One": Handheld technology becomes particularly impressive when you begin using it with your colleagues. This part discusses three food preparation teams who put the handhelds to different uses.

Part 3 "On Being a Project Champion": If you have made it this far, you are probably already your department's expert on, and champion of, handheld technology. This part explains the tools and techniques you will find useful as you introduce the technology to your team.

Begin at the beginning, with Chapter 1. After this, feel free to dip into the other chapters in the order that interests you. We wish you a pleasant read.

About the Authors

Susan comes from Caracas, Venezuela, but headed north to pursue her passion for cooking. She was motivated to become a chef not only because of an interest in food as an ingredient, but also because she considers it a voice of truth. She has held management and administrative positions at Marriott International and PepsiCo, Inc./Pizza Hut, and has worked as a consultant for the Pillsbury Company and Domino Foods. Trained at the Culinary Institute of America, Hendee is certified by the American Culinary Federation as a culinary educator and has served as director and adjunct faculty at New York University's Department of Foods and Nutrition and chef instructor at Johnson & Wales University. In addition to an A.O.S. degree in culinary arts from the CIA, she has earned both a bachelor of science degree in food service education and a master of science degree in computer education from Johnson & Wales, as well as an M.A. in food service management and a Ph.D. in foods and food management from New York University.[2] She continues her celebration of the culinary field by educating future culinarians at the New York Institute of Technology and in her newly appointed position as dean at Baltimore International College. Chef Hendee recommends utilizing a computer (handheld of course), Chinese cleaver, and emotional intelligence[3] as kitchen tools, and a variety of vinegars and sea salts as essential components in great cuisine.

Mohammad is a Bahraini who graduated as a doctor from the University of Cambridge in England. During his medical studies he kept up his interest in computers. In every speciality that he studied, and later, in every hospital that he worked in, he saw clinical problems that could be helped with a little computer technology. After his first year as a resident, he wrote *Handheld Computers for Doctors*. To date, he still receives emails from doctors around the world who are following his advice and using handheld computers in their clinical work.

And it was as a result of reading that book that Susan contacted him and began the partnership for this book.

Acknowledgments

Several people were kind enough to provide feedback on parts of this manuscript. We thank them for their input. They include Sarah R. Labensky, Favorite Recipes® Press, and Francis T. Lynch, Chef Desk.

End Notes

1. Mohammad wants to apologize to *Star Trek* fans for the harshness of his opinion. However, if you would like to read some truly imaginative science fiction, try *Last and First Men*, Olaf Stapledon, ISBN 185798806X.
2. *Measurement of Differences in Emotional Intelligence and Job Satisfaction of Practicing Chefs and Culinary Educators as Measured by the Mayer-Salovey-Caruso Emotional Intelligence Test and Spector's Job Satisfaction Survey.* Proquest—Publication Number 3045714 (September 2002).
3. 2004 International Society Travel and Tourism (Hong Kong, China) and Department of Recreation, Park and Tourism Sciences, Texas A&M University (2004). *A useful extension of previous research that contextualizes the occupation/profession of chef and chef educator within the wider framework of Emotional Intelligence (EI) and its application/relevance to hospitality services.*

HANDHELD COMPUTERS FOR CHEFS

PART 1

WHY STAR TREK

Chapter 1

So, You'd Like a Handheld

"I can't keep track of all the pieces of paper for my job. I think I need a bigger pocket."

No, you don't. You need a PDA.

"Great, another three-letter abbreviation."

PDA stands for personal digital assistant. It is a computer that is small enough to hold in your hand or keep in your pocket. It helps you organize all the information you need to keep track of as a chef or food-service professional. Susan uses hers for prep lists, lists of food stuffs to order, and pictures of plated foods for new menu consultant projects.

"A computer? That means it's expensive, takes ages to figure out, then crashes."

Not necessarily. Many handhelds are quite the opposite. This is why you might have seen a few of your colleagues using them in foodservice. Handhelds are already very popular in corporate and institutional sectors. It is increasingly becoming the norm for independent restaurateurs, waitstaff, and customers to use them, and they're beginning to become popular "tools" with back-of-the-house personnel. If you don't get in the habit of using one, you may be missing out.

"OK then, tell me more."

Handhelds are now affordable, with prices starting at $150 for a good machine. If you want extras such as a camera, Internet access, a telephone, voice recorder, or even a car-navigation system, you will pay more. Nevertheless, even the cheapest model is powerful enough to handle some generic foodservice software.

They are also simple to use. Whereas your first brush with a desktop computer was probably a painful learning experience, you will quickly learn how to use a handheld. You can write on one using a pen (or even your finger), and the programs are simple and clear.

Nor do handhelds crash all the time. Instead, their reliability makes them perfect for the foodservice setting.

"But how does this 'PDA' thing help me in my job?"

All handhelds come with the four essentials—diary, address book, to-do list, and memo pad. These programs are simple and quick to use but much more powerful than any paper organizer. For example, the to-do list lets you keep track of your jobs and arrange them by employee and production and delivery schedules. The diary keeps track of all your appointments. The address book includes the phone numbers of all the employees and vendors for your foodservice operation.

One way to collect all these numbers provides a further illustration of the machine's power—the capability for a colleague to beam them to you.

"'Beam?'"

Beaming is the process of transferring information from one handheld to another using each machine's infrared port. So, when you bump into a friend who has a handheld with the information you need, just line up your machine with theirs, and ask them to beam the information you need. In a few seconds, you can have over 100 addresses copied across. Then you can beam your own addresses back to them, and so on.

"That's very caring, but can I do anything else apart from swapping phone numbers?"

This is where the fun begins. There are literally thousands of programs available to allow your handheld to do extra things. These include recipe references that compile shopping lists; books, music, and videos for your entertainment; the latest news and journals; and games, of course. Many of these take advantage of beaming technology.

"Now you're talking. How do I get one of these machines?"

That's easy—they are on sale in most computer shops, in cell phone shops, and via the Internet. The trouble is choosing from among all the models.

Your first decision is probably determined by your budget. Prices start at a very reasonable $150, but you can easily spend $900, if you are not careful. For $200, you can get a machine, with a color screen, that runs Microsoft Word and Microsoft Excel. After that, extra money gets you cool features but does not affect the kind of programs you can use on it.

Your next choice is between a handheld with the Palm or Pocket PC operating system.

"I knew this was going to get technical. . . . "

All you need to know is that there are currently two major types of handhelds: one works with the Palm operating system, and the other with Pocket PC. Palm machines are the cheapest and simplest to use, whereas Pocket PCs tend to be more powerful. One thing to be aware of though is that many of your colleagues will have Palm machines. That means most generic foodservice software is only available for the Palm, and beaming is easier if you have the same type of device as your colleagues do.

Finally, you should always check before buying the handheld that it works with your main desktop or laptop computer. Why? Because handhelds can place a copy of all their information on a desktop or laptop computer. It is a great feature, so if anything bad happens to your handheld, such as theft or accidental coffee spillage, a copy of your information is still safe.

Try doing that with your pocket full of paper.

Chapter **2**

Choosing Hardware

Jeff Hawkins, the designer of the original Palm Pilot, carved a small piece of wood that was literally the size of his palm and showed it to colleagues at 3Com. This, he announced, was going to be the size of his new organizer. At the time, nobody believed him—Apple was the pioneer of this market, and its organizer, the Newton, was the length of a forearm.

Now, tens of millions of handhelds later, there are many models to choose between. Our advice: decide how much you want to pay, and then find the machine that gives the best features at that price. Otherwise, the sales clerk is likely to dazzle you into unbudgeted territory.

Should You Choose a Palm or Pocket PC?

Palm is not the only company making handhelds. The biggest competitor is Microsoft with its Pocket PCs, but you may find machines by many others such as Blackberry and Sharp.

What it comes down to is a choice of operating system. The operating system determines how much foodservice software you will find and how many of your colleagues can beam data to you. Because of this consideration, you really should not get any handheld that does not have the Palm OS or Microsoft Pocket PC logo. Resist the temptation to get any other kind of handheld. No matter how cheap or impressive it looks in the shop, what software it is possible to run on it is too important a consideration to overlook.

Choosing between Palm and Pocket PC is tricky. The importance of beaming means that an easy rule of thumb is to buy a machine that runs the same software as the machines of your friends and colleagues. You can enlist their help when learning how to use your new handheld computer, and your software is more likely to work smoothly with theirs.

Palm OS logo. You will find this on handhelds made by Alpha-Smart, Garmin, Lenovo, LG, Kyocera, Palm, Samsung, Sony, and Symbol.

POWERED

Often, you will find that your friends get very emotional about which machine you should buy. In fact, this will happen even if you don't ask for their opinion. You should, at least, look on the positive side, that is, people tend to become very emotionally attached to their handhelds.

Perhaps we can offer a more objective view. On the one hand, Palm-compatible machines are cheaper and simpler to use, and have better battery life. On the other hand, Pocket PCs tend to be more powerful. These are general guidelines, not carved-in-stone rules, because all handhelds are improving so quickly.

However, it is useful to clear up two misconceptions. The first concerns popularity. At the time of writing, December 2006, Palm-compatibles are roughly equally popular to Pocket PCs, and have been around a little longer. Therefore, a lot of foodservice software is currently made for the Palm first, and many of your colleagues will have Palm-compatibles. However, you should not rule out a Pocket PC. The machines have gained in popularity, and software developers are responding. Modern Pocket PCs allow beaming with Palm-

Pocket PC logo. You will find this on handhelds made by Acer, Audiovox, Casio, Dell, HP, Motorolla, NEC, Palm, Siemens, and Toshiba.

compatible machines, so you should still be able to work with your colleagues.

The second is about compatibility with your main computer (PC). You do not need a Pocket PC to work with your existing Microsoft software. All Palm-compatibles can work flawlessly with Microsoft Outlook straight out of the box. In addition, all models costing more than $200 include software for viewing and editing your Microsoft Word and Excel documents. In fact, this software works slightly better than a Pocket PC's software does. And for a little extra money, you can buy more powerful versions that allow you to edit your PowerPoint slides on a Palm or Pocket PC.

In conclusion, there are slight differences between the two types of handhelds, but both companies are competing extremely hard for your money. That means you are likely to fall in love with whatever you choose in the end.

Which Palm or Pocket PC Should You Choose?

The most important decision to make is how much RAM (memory) you can afford—this is how much information your handheld can hold and denotes, therefore, how many programs and texts you can keep in it. The next decision is size of screen. With a large screen you can see more, but the device will be more expensive and may be bulkier. Finally, with a camera you can take pictures of your menu items for your team.

If you go for a machine with a lot of RAM, a large screen, and a camera, you can then choose among a host of other features, including telephone capability, a wireless Internet connection, and global positioning system (GPS) navigation. None of these is necessary, but most are highly enjoyable.

Finally, some features are extremely useful for specialized operations. For example, Symbol's devices include barcode scanners and temperature probes, and are ruggedized. The case studies in Part II make extensive use of and discuss these options.

INVENTIVE INPUT

Some handheld computers, especially those with phone capabilities, have a built-in small keyboard. This keyboard works surprisingly well, but the trick is to hold the device with both hands, then use both thumbs to do the typing.

You can write straight onto other handheld computers. This feature is not, however, perfect, and as you grow increasingly reliant on the handheld, you might want to shop around for some extras that will make text input easier.

Two such solutions are styli and keyboards. The first option to consider is Karl Robb's TrueTip finger stylus. It expands to fit on the tip of any finger, but the index finger gives the best control. The idea came to Karl the first time he saw someone using a handheld, and he quickly set about starting his own company (www.truetip.com). *You might want to keep these out of the kitchen, or you may be serving them in someone's soup.*

An alternative model, the pen-cap stylus, fits on the bottom end of most pens. This means that you can write on paper using the pen and then write on your handheld using the pen's cap. Finally, because we constantly lose styli, we particularly appreciate the price of $10 for a four pack.

Bob Olodort of Think Outside took a completely different approach to input. After four years of prototyping, he finally produced the Stowaway unfolding keyboard. This costs about $100 and is invaluable for long writing sessions, including lectures. It is the size of your handheld but unfolds (beautifully) into a full-sized keyboard. This trick will draw gasps from any audience.

TrueTip pen-cap stylus.

The amazing Stowaway keyboard by Think Outside (www.thinkoutside.com).

USE PROTECTION

The kitchen environment is full of grime, grit, and rough and tumble, so it is highly recommended that you invest in protection for your device. For the screen itself, screen shields can prevent scratching or cracking from sharp objects and debris. And hard cases can help your device survive a fall. These and other accessories are available from the website of your device's manufacturer. A specialist site for the sturdiest or "ruggedized" handhelds is Symbol Technologies (www.symbol.com), and it will be discussed further in the Part 2 case studies. If you have disposable income, try PocketSolutions (www.thepocketsolution.com).

Where to Buy Your Handheld?

Most electronics shops in malls sell handheld computers, as do chains such as Best Buy, Circuit City, and Staples. They are, however, generally cheaper to buy online. So, if you are comfortable with buying on the Web, and confident that you have chosen the right model, take advantage of the price differences. Good Internet sites include Amazon (www.amazon.com) and eBay (www.ebay.com). You can also buy directly from the manufacturer, for example, at www.palm.com, www.dell.com, and www.hp.com. Finally, a visit to CNET's handhelds section (www.cnet.com) is always good for comparing prices.

However, if you are buying more than two handhelds, it is often possible to haggle with shops in malls. Susan inherited an HP iPaq and Dell Axim, purchased a PalmOne Zire, and is currently negotiating for a Palm Treo 650 free of charge (more in Chapter 20). Although she does not advise making all these purchases, if someone is willing to upgrade, grab the opportunity to have them share their older handheld with you. The more you talk with people, the more you may find these golden opportunities at no cost to you. It allows for less costly experimentation.

Finally, do not forget that buying a handheld is only half the story— you need to buy additional software that makes it beneficial for a food-service professional. For this, budget at least $100 and the next chapter tells all.

FURTHER INFORMATION

- Amazon—www.amazon.com
- Amazon UK—www.amazon.co.uk
- CNET—www.cnet.com
- Dabs Direct—www.dabs.co.uk
- Palm platform—www.palmsource.com
- Pocket PC platform—www.pocketpc.com
- PocketSolutions—www.thepocketsolution.com.
- Symbol Technologies—www.symbol.com
- Think Outside keyboard—www.thinkoutside.com
- TrueTip stylus—www.truetip.com

Chapter **3**

Choosing Software for Yourself

Computers are only as good as the software that runs them, and handhelds are no exception. In fact, one of the key reasons for buying a Palm-compatible or Pocket PC rather than any other handheld is the range of software available. The chief executive officer of Palm likes to refer to the Palm Economy—a bustling cottage industry of thousands of developers, each contributing their own unique solutions for Palm machines.

Once you have gotten over the initial pleasure of all this software, there soon follows confusion over how to deal with it. The rest of this part of the book will look at the best applications for the most common tasks in foodservice. For those who want to get started immediately, there are three principles to bear in mind.

Find a Guide You Like

In the old days of the Internet, two Stanford students made their Internet bookmarks available to the rest of the world. They were shocked at how popular this list was. In classic California style, however, they recovered and made a company out of it. Yahoo! remains one of the top Internet sites worldwide. Now, whenever a new area of expertise arises on the Internet, it takes very little time for similar bookmarks to arise. These should be your starting points whenever you want to find software.

Our website for non-foodservice software is Handango (www.handango.com). Apart from being the most polished, it is also the most user-friendly. In Yahoo! fashion, it provides a hierarchy of good software. For example, the Document Management category lists the different tools available for working with Microsoft Word, PowerPoint, and

Excel documents. It also has numerous reviews written by people who have bought the software. On other sites, it is quite easy to arrive at a list of over 50 programs, simply because they have not been subclassified. By contrast, Handango's hierarchy is much deeper and usually avoids this situation. For example, the Document Management category includes the Word Processing subcategory, which lists some of the alternatives discussed in Chapter 5.

Websites focusing on information technology needs for chefs are few but growing. None deal with handhelds specifically at the time of this writing, so we created www.handheldsforchefs.com. We focused the website on the needs of chefs and foodservice professionals with reviews by chefs. We also provide an electronic newsletter that you can subscribe to, and a forum for your discussions. Access to the site is free of charge. To *discuss* information technologies in general, three foodservice forums are particularly popular. The first is forums.chef2chef.net, a culinary portal, the second is www.webfoodpros.com/discuss, a forum for food professionals, and the third is www.foodservicei.com/forums, where the foodservice industry checks in. All three are free to join and can provide you with hours of entertainment.

There are other foodservice forums, but they are membership based and do not specifically include information technology, much less handheld talk. The case studies in Part 2 of this book showcase additional professional proprietary and specific vendor resources in detail.

Have a look around these sites, and find ones that you are comfortable with.

There's Lots of Free Stuff Out There. . . .

Many programmers make their work available for free. They do this for the fun and community service, and there is some real quality out there. Even better are the free documents available. Once you have bought document readers or databases, some food enthusiasts make incredibly high-quality content freely available. From recipe databases to diet analysis to food references to the inevitable restaurant reviews—if you need it, someone is likely to have provided it for you. You can gradually build up a library of information you need for your specialty. The best thing about this is how easy it is to share with others—once one team member has downloaded an item of interest, the next team meeting allows beaming to everyone else.

Recipe database

Susan recommends chefs hook up with "techie" friends to brainstorm new foodservice games and creative reference materials. New ideas are sorely needed to include cross-cultural food celebrations, endangered foods, food bioterrorism solutions, Quick Service Restaurant menu changes that are healthy and safe for food allergic customers, and team solutions for day-to-day service problems facing our indus-

Try programs out before purchase

try. We are losing our best and brightest to other creative professions, and the solution does not lie in more recipe databases. Many of our employees and students are the X, Y, and boomer generations, and technology can be a motivator of the young to relate to the old.

Even the Programs That Cost Money Are Free to Try

Do not buy any programs. At least, not immediately. Many handheld programs have at least a 30-day trial. If you do not like the program at the end of that time, just delete it. There is no obligation to pay, and no one will chase you for money. One reason you rarely find negative reviews of handheld software—not having to part with your money makes the whole thing very amicable.

When you do find software you like, do not give your credit card number to any website that is not secure. In other words, if you are using Firefox or Netscape, you should have the following locked icon at the bottom of your window.

And likewise for Internet Explorer, you should see the following icon:

That's it really. Go out there and find the good work. And don't forget that once you start to make handheld content yourself, you too can share it with the Palm Economy.

FURTHER INFORMATION

- Chef2Chef—forums.chef2chef.net
- Handango—www.handango.com
- Handhelds for chefs—www.handheldsforchefs.com
- Foodservice—www.foodservicei.com/forums
- Webfoodpros—www.webfoodpros.com/discuss

Chapter 4

Organizing Your Life

N apoleon was one of history's most successful workaholics, and he expected similarly high standards from his generals. Only a busy man, he would tell them, can have spare time. Depending on your mood, you can look at handhelds as helping you organize your busy work around the foodservice operation or your spare time outside of it. Either way, it would make Napoleon proud. When you buy any handheld, it includes at least four applications. The Palm operating system's designers have named these the Date Book, Address Book, To-Do List, and Memo Pad. Microsoft uses different names for the Pocket PC, but the Calendar, Contacts, Tasks, and Notes are almost identical to their Palm counterparts. For simplicity's sake, we shall use Palm's naming when referring to either platform.

These four basic applications are the heart and soul of your machine. Although several advanced extensions exist, this suite does a surprising amount, while maintaining simplicity.

Date Book or Calendar

Click on the Date Book (Calendar on a Pocket PC) button, and you arrive at your schedule for the day. To add a new event, tap the screen for the time you want it to begin. You can then specify the end time, add an alarm, or "repeat" the event. We found the latter most useful for a classroom teaching schedule or meetings with vendors. This also works well for anniversaries and events, especially when you take advantage of the alarm to remind you two weeks in advance to buy a present. This is great for couples and has kept many a relationship from strife.

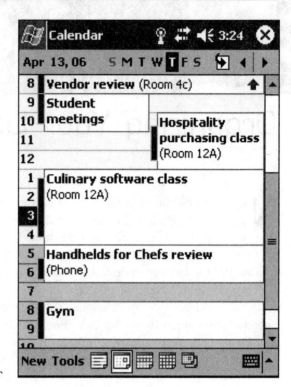

Pocket PC Calendar

Address Book or Contacts

We write everything down in our Address Book (Contacts on a Pocket PC). No scrap of paper escapes this fate: the name of any person whom we meet, contact details of any supplier we deal with, or directions to any place that we visit. It only takes a minute, and it has paid us back immensely. When it comes to organizing any team, Susan knows which times are best for contacting a colleague, a student, a vendor, or an employee.

Your ability to talk to management could improve by leaps and bounds because you can have all their details all the time. Additionally, when arranging externships for your students, you can visit all the food-service sites at a moment's notice, because you have all the directions and quirks in hand.

We do *not* recommend storing your private numbers and passwords in your handheld's Address Book. Many people are tempted to do this. Even when the information is stored in a private address record (one that can only be read when the handheld has been unlocked), it is easy to forget to lock your handheld up at the end of a session. The problem is worse when you think about backing up your data on your main computer—security there is not so good, especially with Palm-compatibles.

We do recommend using a program like eWallet (www.iliumsoft. com) for storing sensitive information like computer passwords and credit card details. It works on Pocket PCs and Palm-compatibles and has foolproof but convenient security.

To-Do Lists or Tasks

With the To-Do List (Tasks on a Pocket PC), you can keep track of all the jobs you have to complete. Of course, these can include the daily activities for any AM or PM shift.

There are also more creative uses for this software. For example, you can store your grocery shopping or that last-minute food request as you pass through the kitchen. Make the category the name of the store or vendor, then assign a date and priority for the item. Next time you are in the store or ordering from a vendor, you will know what to buy. You can also set alarms for items that you have to buy before a certain date.

One other feature to mention is the ability to look at your day's appointments and tasks on one screen. For some of our highly organized yet paper-based colleagues, it was only seeing this view that provided proof of the usability of handheld computers.

Memo or Notes

The Memo (Notes on a Pocket PC) is great for writing scraps of information. Do not underestimate the power of this feature. More expensive devices go even further, allowing you to record speech.

Susan's favorite use of the Memo/Notes feature is to write down fellow chefs' and students' bits of knowledge, "what is new on the market or a new flavor hit," picked up as she passes through the hallways and kitchens. It is then easy to share these with others. During exam time, Mohammad devotes most of his machine to educational snippets.

If you take one thing from this chapter, it should be the importance of the Find function. Use this to search for anything anywhere on your handheld. Consider this simple example. Suppose that, during your

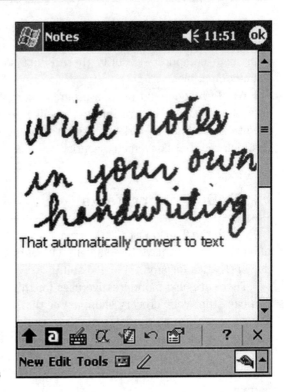

Pocket PC Notes

studying, you wanted to look up guidelines on purchasing wines. If you only had paper notes, you would pray you had stored these notes in a sensible place to quickly find them again. On the handheld, however, you would simply write "wine" in the Find program. Within in a few seconds, this would bring up all the memos you had written about the topic—as long as you had included wine in the text.

But there's more. After a few more seconds, the program would identify any upcoming lectures that you had about the topic. If you waited some more, it would then find the contact details of the instructor who had given the original lecture on wine. The Find program is no slouch—it will search through the whole of your handheld to make sure you can find the answer.

Taking It Further

As we mentioned in the previous chapter, the Palm Economy allows extensions to your handheld in every way imaginable. For example, you can use DayNotez (www.natara.com) as a journal for your class or days tasks, tracking your experiences for your instructor or your employer. It also works as a personal diary, its encryption protecting your secrets from prying eyes. And you can separate the journal from the diary events.

Because the date book and address book programs are so useful, it is worth investing in extra software that combines their strengths. Agendus (www.iambic.com) does this well on Palm Powered devices, whereas Agenda Fusion (www.developerone.com) is designed for Pocket PCs. They allow you to match each appointment in your diary with the contact or contacts from your address book who will be attending the appointment. This is useful in several ways. For example, on the day of an appointment with a vendor, you can quickly tap on the name of the person whom you will be visiting to see their address and any directions you have written for the journey. It is also useful if you have forgotten the name of the supplier but do remember the date of the meeting. Find the meeting, and you will find the name. Finally, you can quickly see the dates on which you had other meetings with the vendor because those events are automatically listed by the software under the person's contact details.

It will not take you long to find needs that your machine does not meet by default, but you will no doubt enjoy the search for the solutions. Happy hunting.

 FURTHER INFORMATION

- Agenda Fusion—www.developerone.com
- Agendus—www.iambic.com
- DayNotez—www.natara.com
- DualDate—www.palm.com/support/dualdate
- eWallet—www.iliumsoft.com

Chapter 5

Taking Notes

Lectures are the process by which a speaker's notes become the student's notes, without passing through either's brain. With a handheld, it is now easy to transfer these notes to other students, also bypassing their brain. As always, you have a wide choice of ways.

Memo or Notes

Memo is great for little scribbles. As mentioned in Chapter 4, Susan uses the program for capturing kitchen conversations, and Mohammad used the program for exams. Susan now has a list of dos and don'ts on her Palm, ready to beam to students at the end of her lectures. Students find it useful to refer to such lists while in the lab. Their learning curve improves rapidly after this because there is no longer a separation between what they needed to do and what they remembered to do. Best of all, Memo comes free with your handheld.

Nevertheless, it does have its limitations. The length of each note is only a few screens, which will not be sufficient space for an entire lecture. Moreover, the closest thing to a heading that is possible is writing in capitals.

Word Processors

Pocket PC users have a program called Pocket Word. Both this and Pocket Excel are included free with every machine. Most Palm-compatible machines also come with equivalent software, called Documents To Go (www.dataviz.com).

For a little more money, you can extend these programs to edit your PowerPoint documents. Palm users also have a range of more powerful alternatives for their Microsoft Office documents. In fact, some of

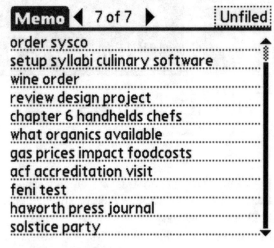

order sysco
setup syllabi culinary software
wine order
review design project
chapter 6 handhelds chefs
what organics available
gas prices impact foodcosts
acf accreditation visit
feni test
haworth press journal
solstice party

Memos and notes help   (Done) (Details)
keep you organized

these, like Quickoffice Pro (www.cesinc.com), are better at dealing with
Microsoft Word documents than Microsoft's Pocket Word. This is why
in Chapter 2 we said that you do not need a Pocket PC to ensure Microsoft compatibility.

If you regularly create documents in Microsoft Office (Word or Excel), then these are the programs for you. Their integration with both

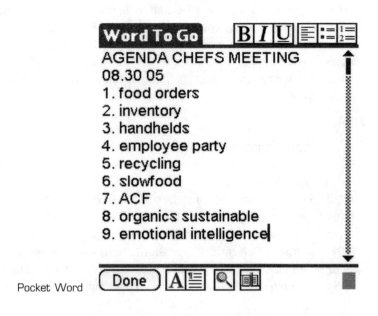

Word To Go   B I U

AGENDA CHEFS MEETING
08.30 05
1. food orders
2. inventory
3. handhelds
4. employee party
5. recycling
6. slowfood
7. ACF
8. organics sustainable
9. emotional intelligence

Pocket Word   (Done) A

	A	B	C	D	E	
1	Name	Information				
2	PrimaryS	11-a				
3	Supplier	MyGrocer				
4	Supplier	100 Main Street				
5	Supplier	Any City				
6	Supplier	Any State				
7	Postal Co	00000				
8	Country	Any Country				
9	Phone	000-000-000				
10	Fax	000-000-000				
11	E-mail	anybody@mygrocer.com				
12	Homepa	www.mygrocer.com#http://www.mygrocel				
13						

Pocket Excel

of these Office programs is powerful and professional, and the display is a delight. Just look at the pictures above.

Mohammad's wife, Laura, used this software to store her notes during college. She loved having a word processor on her handheld, as do many millions of other users. However, linear notes are not to Mohammad's taste. Instead, he uses outliners.

Outliners

If you like the outliner tool on Microsoft Word, then Thought-Manager (www.handshigh.com) is the program for you. It works in the same way as in Word and is an extremely elegant, structured, and compact way of making notes.

Most people have not used an outliner, and they initially find this way of note taking strange. However, if you have ever learned about sauces, you have likely picked up the outliner way of thinking, even though you might not know it.

Do you remember learning to classify all the leading sauces? Do you remember dividing them into béchamel, velouté, espagnole, or brown, tomato, and hollandaise, then subdividing béchamel derivatives into Dill, Duchess, Mustard, and more; and then hollandaise into Maltese, Marguery, and Omega? Most students go through a eureka moment as they realize how this system simplifies the learning of so many lists.

Start with the heading

Expanding the leading sauces

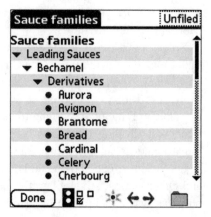

Expand all the way

Outliners make use of this power of classification. The screenshots opposite are taken from Chef Bruce Spivak's lectures on sauces. The structure of the document helps students learn the details. The first screenshot also shows an extra advantage—you can see all the major headings on one screen. Given the small screen of some handhelds, this is a great feature. This is why Mohammad chose this method for writing all his lecture notes. You must decide for yourself. As with all other handheld software, ThoughtManager comes with a 30-day trial period, so you can try it before opening your wallet.

For Susan, voice recording is another great tool.

Voice Recording

High-end devices include voice-recording features. Susan uses this for self-reminders to place orders, contact staff and students, and follow up research ideas. With extra storage cards she can also record entire interviews, and menu and staff meetings, which she then transcribes using a keyboard.

Joy Machine

It is worth remembering how useful a keyboard is for speeding up note taking. And that the true aim of gadgetry is a feeling of smug superiority. So, when the lights dim for the overhead projector, and your friends abandon their pens and paper . . . switch on your machine's backlight.

FURTHER INFORMATION

- Documents To Go—www.dataviz.com
- Quickoffice Pro—www.cesinc.com
- ThoughtManager—www.handshigh.com

Chapter 6

Taking Your Media with You

Picture the scene. Susan is listening to her favorite Tina Turner album. She is reading her student's multiple foodservice projects. After reading each report, she adds notes about the student in her database.

All this is happening on the beach at Fire Island, and Susan's only tools are a handheld computer and a pair of sunglasses. No folders, no CDs, no paper. Life is sweet.

RepliGo and MobiPocket

As the previous chapter showed, it is easy to read and edit Microsoft Office documents on a handheld computer. Sometimes, however, you just want to read the documents, without editing them. Software like RepliGo (www.repligo.com) and MobiPocket (www.mobipocket.com) is designed for reading. This means that fewer pixels on the screen are used up for editing tools, and more space is devoted to the text you want to read, and the rendition of each letter is richer and easier on the eye.

These tools are especially important if you want to share the files with colleagues. For example, Susan likes to beam her class notes to her students so that they can beam them to other students, but it is important that students not be able to change the files, so that they do not spread incorrect information.

There are two versions of RepliGo and MobiPocket. The creator version—RepliGo or MobiPocket Publisher—converts your Microsoft Office document into RepliGo or MobiPocket format, respectively. In fact, RepliGo can convert any document that can be printed. So, if you have recipe software on your PC that prints individual recipes, you can use RepliGo to store a copy of those printouts on your handheld computer. These versions of the software cost around $30 each.

But the reader version—RepliGo Viewer or MobiPocket Reader—is freely available. That means you can buy a copy of the creator version for your own PC, but your colleagues can get free copies of the reader software to use the files that you convert for their handheld computers.

The RepliGo Viewer has one particularly nice feature called text reflow (see parts c and d of the figure below). It means that you can switch from seeing a page as it was meant to be printed—with its

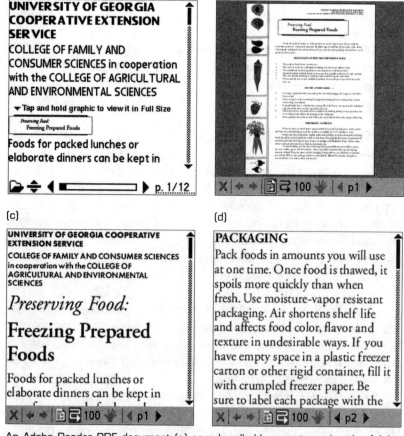

An Adobe Reader PDF document (a) on a handheld computer using the Adobe Reader software. The other screenshots (b, c, and d) are from the RepliGo Viewer software. The bottom two screenshots are in the text reflow mode.

columns and pictures laid out as they would be on paper—to seeing it optimized for the width of your screen. In the latter mode, no matter how wide the original columns of text were, the RepliGo Viewer reflows the text so that you never have to scroll right or left. You can simply keep your thumb on the down button of your handheld computer to read more of the document, making for a comfortable experience.

In fact the RepliGo Viewer is even better for reading Adobe Reader documents than using the Adobe Reader software on your handheld computer (see page 44).

To convert a document on your computer to RepliGo format, open it and select Print from the file menu. Select RepliGo as your printer, and RepliGo will create a version for your handheld computer. It will appear there the next time you synchronize your computer. This is how Susan could convert her student's projects for her handheld computer in preparation for the beach.

The MobiPocket Creator only converts Microsoft Word documents, Adobe Reader PDF documents, and HTML (i.e., web page) documents. This means that you cannot carry your PowerPoint presentations. Furthermore, the PDF conversion is not as good as RepliGo. On the other hand, if you have bought books and textbooks for your MobiPocket Reader software you might prefer to just use that software to read your Word reference documents.

Databases for Storing Recipes

Susan could have made notes about her students in the Memo or Notes software on her handheld computer. Or she could have created a Word document on the device for formatting her scoring. However, in this case a database is better.

A database is like an electronic filing cabinet. Like the paper forms you fill out before putting them in the cabinet, it allows you to enter information about something. And like the filing cabinet it allows you to store all that information in one safe place.

However, unlike a paper form, a database's form can speed up the entry of information, for example by automatically including the date on which Susan is marking the projects. And unlike a filing cabinet, a database quickly sorts and resorts the data in any way you want it, and provides sophisticated search tools so that you find the subset of information that you need. This means that Susan can order her notes by the last name of the student or by the scores. And she can find the students who scored above a certain mark.

Several database tools exist, but HanDBase (www.ddhsoftware. com) is the best for most people. First, its interface is powerful, with sophisticated design tools viewable on the large-screen PC version. Second, both Pocket PC and Palm Powered versions exist, and they include encryption to keep your data secure. This way Susan can keep her scores secret. Finally, the software has a peer-to-peer synchronization feature, which means that two handheld computers can compare each other's records and exchange any missing data. For example, students who store their recipes on their handheld computer can have the software identify recipes in their colleagues' devices that they lack, and then copy those over (with their colleagues' permission).

Filling out forms on a HanDBase database is fairly intuitive, but the company's website (www.ddhsoftware.com/support) includes some useful video tutorials that quickly teach you. The videos can also teach you how to design your own database, but if you have never done so it is often easier to change someone else's design. The HanDBase gallery (www.ddhsoftware.com/gallery) has over 2000 examples that you can freely download, including a "Food and Drink" category.

Music

And then there is the music. If you have music on your own CDs, or in MP3 format,[1] then you can play it on your handheld computer. Of course, you should only do this with music that you have paid for—either by buying the CD or by paying for the MP3 tracks—so that the artists who created the music get their fair share and so that you stay within the law. If your track is already in MP3 format, then you can drag it onto your Pocket PC for synchronization or add it through the Palm Quick Install tool for Palm Powered devices. If it is on a CD, you can convert to MP3 format using your existing software, that is, Windows Media on Windows PCs and iTunes on Macs and Windows PCs. Alternatively, Zlurp (www.zlurp.com) is freely available open source software that is more specialized and creates higher-quality MP3 files.

Any handheld computer that comes with speakers (i.e., almost every one) comes with software to play the MP3 tracks. On Pocket PCs this is usually the Windows Media Player, whereas on Palm Powered devices the Real Mobile Player is common. However, their features are limited. The Core Media Player is open source and freely available at www.tcmp.org. It can also show photographs and play movie files. If you would like to manage your playlists, for example by arranging the order in which your tracks are played, then Busker (www.electricpocket.com) and Pocket Tunes (www.pocket-tunes.com) are good for Palm Powered devices, and Pocket Music (www.pocketmind.com) is powerful software for Pocket PCs.

And that completes this chapter's recipe for beach bliss. Sit back and enjoy the music.

FURTHER INFORMATION

- Busker—www.electricpocket.com
- HanDBase—www.ddhsoftware.com
- HanDBase gallery—www.ddhsoftware.com/gallery
- HanDBase tutorials—www.ddhsoftware.com/support
- MobiPocket—www.mobipocket.com
- Pocket Music—www.pocketmind.com
- Pocket Tunes—www.pocket-tunes.com
- RepliGo—www.repligo.com
- The Core Media Player—www.tcmp.org
- Zlurp—www.zlurp.com

END NOTES

1. Some handheld computers can play music in some other formats, such as Windows Media or Apple iTunes tracks that you buy. However, these are digitally locked to one device and prove rather complex to move around. It is often simpler to just make MP3 copies of your own CDs.

Chapter **7**

Food References and Software

Mohammad once annoyed an instructor by using his handheld in front of him. The instructor had asked him to wait outside his classroom because he was dealing with another student. Mohammad's short attention span soon had him rooting for his device, and he began reading for more information about an assignment. The instructor saw him and suggested that he do some work. Mohammad replied that he was, but the instructor just thrust an article in his direction. Mohammad thanked him for this, and said he would read it as soon as he had finished the chapter. Two minutes later, the instructor was back, and more annoyed that Mohammad was "not doing work." He was now presented with another article. It was that article that broke Mohammad, when the instructor asked him why he insisted on this. He said the school was no place for a "Game Boy," and that Mohammad should be doing some work.

The professor calmed down after seeing the book chapter on the handheld computer.

The Free Stuff

Some good foodservice references are available free of charge, but not many. From Spanish food translations to food restaurant databases to recipe databases, you can find them all on the handheld portals mentioned in Chapter 3. But for the most part if you want quality you have to pay for it.

As always, if you are not sure, then just download it and try it. And if you cannot find what you need already available, consider doing it yourself and making it available to others. Mohammad found HanDBase (www.ddhsoftware.com) most useful for this purpose, but everyone has their own preferences.

Beer Lovers

For the keg connoisseur Palm Brew (http://choiceway.com/palmbrew) contains great information about beer. This includes food pairings, and a directory of brewers and styles of beer. The random trivia section explains the origin of the word *beer* and the role of the Egyptian god Osiris in the creation of beer. The Pocket PC equivalent, from the same company, is Pocket Brew.

The Spice Is Right

Spice used to be expensive. In the story of Genesis, Joseph was sold into slavery by his brothers to spice merchants. Spices were the primary reason that Portuguese navigator Vasco Da Gama sailed to India. And the American continents were discovered when Columbus misnavigated while trying to discover another route to Eastern spices.

The Pocket Brew

Amy Reiley's Pocket Vineyard

 My Tasting Notes

 Glossary

 Vintage Chart

 Food & Wine Pairing

Amy's Selections

The Pocket Vineyard

Fortunately, $9.95 today gets you the spice-o-pedia for your Palm, Pocket PC, or Symbian device (www.l3solutions.com). For over 200 herbs and spices, it lists culinary uses, preparation and storage information, and detailed tips on which spices go best with which foods.

Another $14.95 gets you MyCookbook, available from the same website. It includes a small selection of recipes, each with a general information section, ingredients list, detailed cooking instructions, and optional comments and nutrition information. The power of the software is in allowing you to add your own recipes.

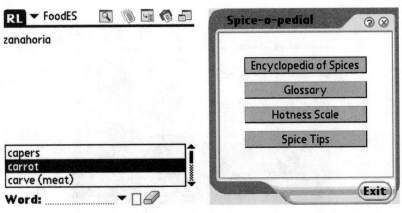

Database of ingredients   Spice-o-pedia

Amy Reiley's Pocket Gourmet

Menu Assistant
2,300+ menu definitions and descriptions.

Restaurant Notes
Record details of your favorite restaurants.

Tip Calculator

Recipe and menu planning

For value for money in the number of recipes and applications, Susan uses MasterCook (www.valusoft.com) with her students. It comes with over 7000 recipes for your PC, and you can select which ones to store on your Palm or Pocket PC. This means that you can provide your kitchen staff with the ingredients, directions, and nutritional information about all the items on your menu.

The Pocket Bartender

Getting Serious

For more professional budgets there is a range of software to meet your needs. For example, The *Book of Yields* (www.wiley.com/college/lynch) can save you weeks of kitchen testing for $80.

*Chef*Tec (www.culinarysoftware.com) provides not only sophisticated recipe management but also inventory taking and ordering capabilities. Each of these three modules costs $595. Its *Cor*Tec prices start at $995. The company is one of the oldest in this price range for chefs and has a proven track record.

The CookenPro Commercial Suite (www.cooken.com) includes a PDA Kitchen Magic module that exports your inventory, invoices, purchase orders, menus, recipes, and other information to your handheld computer.

Finally, the case studies discuss software in the highest prices ranges, namely the products of CBORD, Micros, and NextStep Edge Technologies.

Susan smiles as this chapter ends, remembering the countless calls to Francis Lynch, author of *The Book of Yields* (Wiley 2004) and originator of ChefDesk, and George Kotcher at Barrington (CookenPro). Chefs, we said, will embrace technology and food yields, but the means must be easy to use and in your hand!

FURTHER INFORMATION

- *The Book of Yields*—www.wiley.com/college/lynch
- ChefTec—www.culinarysoftware.com
- CookenPro—www.cooken.com
- HanDBase—www.ddhsoftware.com
- MasterCook—www.valusoft.com
- PalmBrew and PDABrew—http://choiceway.com/palmbrew
- Spice-o-pedia—www.l3solutions.com

Chapter **8**

Reading Electronic Books

An entrepreneur wrote a business plan to create a magazine for insomniacs. The genius of the idea was in advertising on TV at night, when advertising charges are low and insomniacs are many. As one such sufferer, Mohammad needs reading material at night. Even when shattered after a long shift, he cannot sleep without reading. There are many chefs who strongly agree. You know what is coming next. Your handheld computer offers a solution to this problem. Using the backlight, you can switch off the room's light sources and get comfortable with an electronic book. This experience is ten times better if you have a color machine and is another reason we had recommended getting one.

The Pleasure of Reading

Peanut Press was the original name for eReader.com. The company publishes both current bestsellers and classics and a key ingredient of their success is the eReader software (www.ereader.com). Originally known as the Palm Reader, this freely available program allows you to read their books on Palm Powered, Pocket PC and Symbian handheld computers. It also works on Windows or Mac computers. Reading on a small screen differs from reading paper, and the software's designers took account of this. You can change the font size, add bookmarks or annotations, and, best of all, the book automatically returns to the page you were last at. This is extremely important for the continuous interruption of kitchens.

What of the books themselves? Every week, scores are added as the company and readership grow. There is something for everyone. For example, searching for "food" brings up Reichel's *Comfort Me with Apples* and Mayes' *Bella Tuscany*.

6

ARMADILLOS IN CHINA

"The two most important things in life," my ailing father said, "are imagination and laughter. They make anything interesting. Even this."

He gestured at the tray before him, on which was arrayed an extremely dry piece of a substance reputed to be fish, some plain strands of unadorned

462

Over 2000 titles are available in the classics section. This is great for improving your cultural quotient, especially because these books are more reasonably priced.

Which brings us to price. As a rough guide, the cost is similar to Amazon's prices (with its reductions), but without the delivery charges. That means if a book is currently available in hardback, you will pay hardback prices for an electronic version. The pricing is particularly silly for textbooks.

Textbooks

Few textbooks are available in eReader or Palm Reader format. Instead, you will need the MobiPocket reader. It is also available free of charge (www.mobipocket.com), although it does not provide as smooth a read. The MobiPocket website includes several textbooks, but

5

Caviar for the Masses

That is caviar, she explained to him, and this is vodka, the drink of the people, but I think you will find the two are admirably suited to each other.

— C. S. FORESTER, COMMODORE HORNBLOWER

466

you will probably find the ones you are looking for at the eBooks website (www.ebooks.com).

For example, under the "Food and Wine" category, you will find *Food and Cultural Studies* by Hollows et al., and *The Bialy Eaters* by Sheraton.

Most of the books are available in several formats, so look out for the "MobiPocket Reader" phrase or the [icon] icon. Avoid the Adobe Reader and Microsoft Reader formats because of their approach to digital rights management.

Digital Rights Mismanagement

When you buy a paper book, you can resell it to someone else, lend it to colleagues, or photocopy a few of its pages. Electronic books are more complicated. You cannot resell nor lend your copy to anyone else. Some publishers even prevent you from copying a single paragraph.

Problems arise from how they prevent you from reselling, lending, and copying—they do not just use the law, instead they enforce these restrictions using the reader software's digital rights management (DRM). This enforcement can be badly executed.

For example, you can take your paper book from your home to your office, but the enforcement of DRM can stop you from moving your electronic book from your home computer to your work computer. It can even complicate the process of moving the electronic book from your computer to your handheld computer.

Adobe Reader and Microsoft Reader are particularly unpleasant. Many customers have paid for books that are no longer accessible because of the complexity of the reader software.

By far the best software is the eReader or Palm Reader. When unlocking a book, you need the details of the credit card that was previously used to buy it. It is easy to access old books because of this and to use them on multiple machines. The MobiPocket Reader is not quite so lenient, but it is designed with multiple-machine use in mind, and the process is reasonably simple.

Of course, not all books cost money or have DRM.

Free Books

Project Gutenberg makes thousands of books available free of charge, with no DRM restrictions. Most of these are old books for which the copyright has expired and are now available in the public domain.

Project Gutenberg shows the beauty of teamwork. Around the world, volunteers enter the text of these old books using the website's tools. Some do so by typing, others by scanning. The website allows several people to cooperate on the data entry for each book and for others to proofread what has been entered.

The results are available to download at www.gutenberg.org. Examples include *The Food of the Gods and How It Came to Earth* by H. G. Wells and *Tender Buttons* by Gertrude Stein.

There are several ways to read these books because most are available as simple text files. This means that you can read and edit them in WordToGo or Pocket Word (see Chapter 5). However, these software tools are not designed for easy reading, and you may shy away from the ability to edit the words of Mr. Wells or Ms. Stein. Instead, you should use RepliGo (see Chapter 6) or Plucker. As you will learn in Chapter 10, Plucker makes the task particularly easy.

Audio Books

Susan's favorite kitchen veteran, Anthony Bourdain, writes truthfully, but hearing him read *Kitchen Confidential* brings an extra level of enjoyment. He brings culinary stories to life, pronouncing insults as they ought to be.

His performance, along with thousands of other audio books, is available through the excellent Audible (www.audible.com).

Prices initially appear as high as those of hardback books. However, it is worth signing up for the AudibleListener program. For $15 per month, you can choose any audio book, and one month's episodes of any radio program, while $20 per month allows you two audio books.

You can stop paying whenever you want to and restart when you are ready to listen to more audio books. Alternatively, if you sign up for the one-year contract, you can receive $100 towards the cost of a handheld computer, making it a great gift.

The best thing about Audible is its mature approach to DRM. You can place your books on as many computers, handheld computers, and CDs as you need to. In other words, even if you bought an audio book for your handheld computer, you can also install it onto the handheld computer of your partner (or anyone else you trust with your credit card details). Very civilized.

Whatever you decide to use, electronic books are useful to students, professionals, and casual readers. And, of course, there are the insomniacs with backlit screens.

FURTHER INFORMATION

- Audible—www.audible.com
- Ebooks.com—www.ebooks.com
- EReader.com—www.ereader.com
- Mobipocket—www.mobipocket.com
- Plucker—www.plkr.org
- Project Gutenberg—www.gutenberg.org

Chapter 9

Games for Chefs

The instructor who told Mohammad off for playing with his "Game Boy" was impressed to discover that it was actually a handheld with reference textbooks. However, it was a lucky escape. Mohammad's device has its fair share of games, which he has put to use on plenty of other occasions. So, for all those illicit moments in the classroom or kitchen, here is a list of our favorites.

Nutritive Naughtiness

We begin on a culinary theme with three simple games. Snack and Snake (www.allmobileworld.com) is a Pocket PC version of the snake game some of us may remember playing with early computer game systems. You must feed the snake without colliding with the walls, despite the increase in its length with each snack. Brilliance (www.rampart games.com) is a modern version of snap, in which you must remember to match food items against the clock.

Brilliance

Words Hound

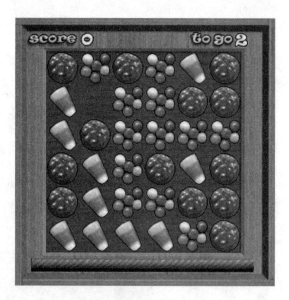

Match the candy

Finally, SnackDicer (www.caproject.com) has an educational claim, teaching you about healthy lifestyle and balanced diet.

Traditional Games

But enough about work: you want to play in those idle moments. Susan's partner Loring's favorite games remain the card variants of solitaire, and Paragon (www.penreader.com) has one of the largest collections. The Pocket PC version is called the Can't Stop Solitaires Collection, while the Palm Powered version is the One for All Solitaires collection.

A friend of Mohammad's was hooked on chess. While working, he would mull over his next move, and each game would take several days. Infuriatingly, as soon as he made his carefully considered move, the computer would immediately respond. He rarely won, but that did not stop him.

There is a free open source version available for Palm Powered devices at palmopenchess.sourceforge.net. In fact, www.palmopensource. com contains a whole range of open source games (and plenty of other software) for these devices, while ppc.palmopensource.com has the Pocket PC versions. At the time of writing, there was no open source version for these devices, but Valentin Iliescu (www.valil.com) has created an excellent version that he makes freely available. It is even suitable for playing with other people across the Internet.

Strategy Games

For Mohammad, however, nothing beats a good game of Civilization. There is no official version yet available, but Kingdom (www.grogsoft. com) for the Palm is well worth the price, while Pocket Humanity (www.pockethumanity.com) is freely available for Pocket PCs.

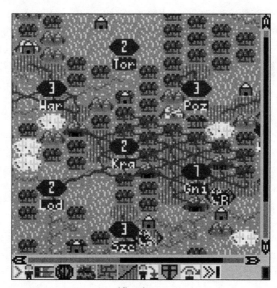

Kingdom

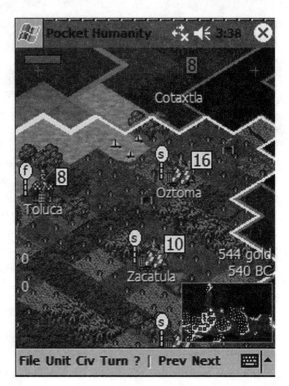

Pocket Humanity

Technicolor Glory

If you're still waiting for adrenaline to flow, do not worry, for there are plenty of games that provide this. One of the most impressive to emerge is Warfare Incorporated (www.warfareincorporated.com). You must guide your futuristic army against the enemy and the interface is sumptuously colorful.

For some hand-to-hand combat, one French developer (yoyofr.fr.st) has created Palm Powered versions of Doom, Quake, and Hexen to satisfy his evident bloodlust. A Pocket PC equivalent is Red Sector 2112 (www.4pockets.com).

Finally, Race Fever (www.digital-fiction.com) is a low-powered car-racing game. But its beauty is that you can play against your friend through the infrared beam of your Palm Powered devices. You may find this addictive. Do be warned, however, that it is not quite as easy to pretend that you are doing work when playing this: the sight of two students struggling to stay within beaming distance while shouting abuse at each other fools no one.

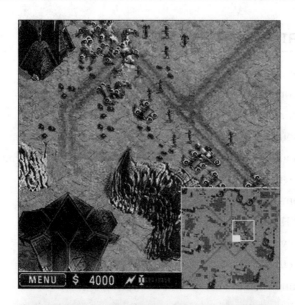

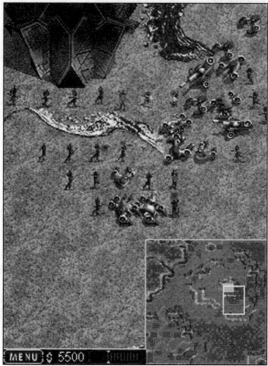

Warfare Incorporated for Palm and Pocket PC.

FURTHER INFORMATION

- Brilliance—www.rampartgames.com
- Chess for Palm—palmopenchess.sourceforge.net
- Chess for Pocket PC—www.valil.com
- Open source games for Palm—www.palmopensource.com
- Open source games for Pocket PC—ppc.palmopensource.com
- Doom, Quake, and Hexen for Palm—yoyo.fr.st
- Kingdom—www.grogsoft.com
- Pocket Humanity—www.pockethumanity.com
- Race Fever—www.digital-fiction.com
- Red Sector 2112—www.4pockets.com
- Snack and Snake—www.allmobileworld.com
- SnackDicer—www.caproject.com
- Solitaires—www.penreader.com
- Warfare Incorporated—www.warfareincorporated.com

Chapter 10

Carrying the Web with You

In 1712, the British government began taxing newspapers based on the number of their pages. Many newspaper publishers responded by increasing the size of each page to reduce the number of pages necessary for each issue. Three centuries later, *The New York Times* and other American newspapers still use that same large size despite the inconvenience to their readers. It is only recently that, especially around metro stations, smaller-sized "express" newspapers are appearing.

As the owner of a PDA, however, you can enjoy your favorite newspapers in the most convenient size.

Introducing AvantGo

On the face of it, AvantGo (www.avantgo.com) is a simple idea—it keeps a copy on your handheld of websites that you like. Later, you can read as much or as little of those sites, without needing an Internet connection. But the company has done a lot of work to extend this. For you and me, they have made the process of acquiring these pages beautifully simple. For the website designers, they have provided tools and advice on how to design sites specifically for small machines.

AvantGo Channels

When you visit the AvantGo website, there is a simple sign-up procedure. It takes you through the process step by step as you download the software for your handheld, install it, and then pick the information you want to take with you. They call this information "channels." Then, every time you backup your handheld onto your PC (i.e.,

The New York Times
ON THE WEB

Sunday, November 20, 2005

Today's Books News

Network Error
By JONATHAN ALTER
From a former producer at "60 Minutes Wednesday," a
high-spirited, if self-serving, account of how a report on
George W. Bush and the National Guard blackened the CBS
eye.

Yes, Virginia
By CURTIS SITTENFELD
Another Woolf biography, this one focused on the work, not
the gossip.

Normal Life With More Pancakes
By JENNIFER SENIOR
A journalist who covered her biggest story: her own death
from cancer at age 47.

Soul Man
By JOHN LELAND
Peter Guralnick's biography about Sam Cooke struggling to

The New York Times Books AvantGo channel

synchronize it), AvantGo checks to see if you have an Internet connection open. If you do, it picks up the latest updates to the channels
you had asked for.

It is always a surprise what you find when you explore the list of
channels, and there are more cropping up everyday.

Start with the Food & Drink subcategory inside the Lifestyle category. Its channels include "Chopstix recipe of the day," which has a
Chinese or Asian recipe every day from www.chopstix.com, and
DrinkBoy, with cocktail recipes from www.drinkboy.com, and all sorts
of recipes at the channel of allrecipes.com. Once you find the channel
that you want, click Add Channel to My Device.

Alternatively, if you were at the DrinkBoy website, for example,
you would have seen the AvantGo logo **AvantGo** inviting you to
subscribe.

You can find this logo on several news websites, including those of
the BBC (news.bbc.co.uk), CNN (www.cnn.com), and of course, *The
New York Times* (www.nytimes.com).

Steamed Scallop (November 19, 2005)
This is one of the best ways to eat scallop. At
Chinese Experience in London, chef Gun Leung
steams...

Kowloon Bakery, London (November 15, 2005)
Kowloon Bakery is one of my favourite places in
London's chinatown but you won't find it in your
fancy...

Quicklinks for Sunday, 13 November 2005
(November 13, 2005)
The Breath of a Wok (Alternate UK link) has
been named joint best food book in the 2005
World Food...

Stock - the real secret of Chinese cooking
(November 08, 2005)
Nothing tells me more about a Chinese
restaurant than its use of stock. Stock is
central to Chinese cuisine....

So how did I get here? (November 03, 2005)

Chopstix AvantGo channel

Vindigo

The restaurant, nightlife, and film reviews of *The New York Times* are available through a different service, Vindigo (www.vindigo.com). For around $25 per year the company's software allows you to download as many city maps as you would like. The maps include most of the large cities in America (even Des Moines and Tulsa), some counties (e.g., Montgomery County in Maryland), and London, England.

The maps show you where you are in two ways. If you have a Global Positioning System device, as some of the more expensive handheld computers do, Vindigo will pick up the signal from the device and center the map on the correct location. Most people do not have this, however, and it is not necessary to make full use of Vindigo. Walk to the next intersection, enter the name of the two intersecting streets into the software. The map will center correctly.

The software can also show you to the route to follow from your current location to where you want to be. Enter the street address of your destination, and VindiGo will resize the map to show walking and

The *New York Times*
on Vindigo

driving directions, as well as telling you which metro stops are next to you and your destination.

The beauty of the software is its ability to help you decide where to visit. With each synchronization, the Vindigo software downloads the latest maps, as well as the weather reports, restaurant listings, movie times, and nearest ATMs. When in a new part of town, you can see which restaurants are closest to you, read their reviews from *The New York Times* or *Zagat*, and see how to walk to there. Susan uses both of these extensively for restaurants in New York City and Baltimore; and *Zagat's* does include Long Island, also serving local needs. And for aging populations VindiGo has, yes, bathrooms!

Incidentally, if your own restaurant is not listed, make sure you contact Vindigo. Handheld computer owners are lucrative customers, and it is important that they know about your restaurant.

Browsing the Web

Vindigo and AvantGo are companies that partner with other companies to bring their websites to your handheld computer. As you rely more and more on your device, however, you will want to look at other content not available through Vindigo or AvantGo's software. This is especially the case with professional culinary websites because they have niche audiences.

Zagat's

Plucker (www.plkr.org) allows you to store entire websites onto your handheld computer. The software uses your PC's connection to the Internet to download the copy of the website to your handheld computer. As with Vindigo and AvantGo, this means that you do not need your handheld computer to connect to the Internet to look at content.

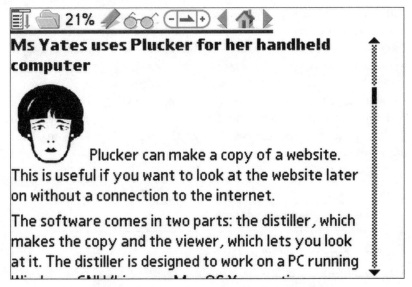

Plucker allows you to download entire websites

For example, you can download the entire text of Mohammad's book *Free Software for Busy People* (see the preceding figure) from www. freedomsoftware.info. You can also download any of the Project Gutenberg (www.gutenberg.org) books mentioned in Chapter 8. And, of course, you can download useful culinary websites such as www.escoffier.com.

However, many of those websites are too large to download in their entirety, or you may want to look at the latest content in their discussion areas. For this, you will need to connect your handheld computer to the Internet.

If your device has a built-in Wi-Fi card and your institution has a Wi-Fi router, you can connect easily and cheaply. If not, you will need to connect using your wireless cell phone company. This is much easier to arrange than using Wi-Fi but also quite expensive. Talk to your wireless carrier about how much they bill for this service; the cost is always dropping.

Both Palm Powered and Pocket PC devices include web browser software to allow you to visit your favorite culinary websites. Web Pro, on Palm devices, is particularly good if you select Handheld View from the Options menu. But the Pocket PC's Pocket Internet Explorer is rather low quality. Fortunately, several alternatives exist.

Minimo (www.mozilla.org/projects/minimo) is made by the same people who make Firefox, possibly the best web browser for PCs. Like Firefox, it is open source and available free of charge. For around $30, you can get the Pocket PC Netfront browser (www.access.co.jp/ english), which is made by the same company that created Sony's Palm Powered web browser. Finally, Bitstream takes a different approach with its Thunderhawk browser (www.bitstream.com/wireless). For an annual subscription of $50, the company's computers will compress each web page for your Pocket PC. This allows quick loading of pages, preserving the original layout but with good legibility on the small screen because of the company's font technology.

With all these tools, you can read the news websites of your choice without the inconvenience of broadsheet newspapers. That means less folding of newspapers, decreased paper consumption, and more pleasant reading.

FURTHER INFORMATION

- AvantGo—www.avantgo.com
- Bitstream—www.bitstream.com/wireless

- *Free Software for Busy People* by Mohammad Al-Ubaydli. Idiopathic Publishing, 2005, www.freedomsoftware.info
- Netfront—www.access.co.jp/english
- Minimo—www.mozilla.org/projects/minimo/
- Plucker—www.plkr.org
- Vindigo—www.vindigo.com

Chapter 11

Handhelds for Food Lovers

The pioneer of tourist guidebooks was Germany's Karl Baedeker, who released a guide to Coblenz in 1829. His rating system was severe—most sites received no stars, and the top mark of two stars was reserved for the likes of the Louvre, the Pyramids, and Yellowstone Park—and he considered overtipping a cardinal sin.

Since then, guidebooks have evolved for every taste and medium for today's lovers of food and wine.

Restaurant Guides

For example, Zagat (www.zagat.com) makes its reviews and ratings available through the products of other companies, including Vindigo's software (www.vindigo.com), which was mentioned in the previous chapter. This costs $25 per year and provides weather reports, movie listings, and directions to the each restaurant along with reviews.

Handmark's Zagat To Go (www.handmark.com) costs and provides similar, but not identical, features. For example, it includes guides to Tokyo but not the weather reports or movie listings.

At the Restaurant

The most interesting foods often have the strangest names—names that are not translated into English in some restaurants.

Good food deserves good wine, of course, and LandWare (www.landware.com) publishes a new edition of the *Wine Enthusiast Guide* every year for around $20.

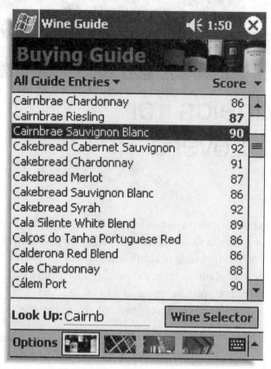

Illustration 11-1 *The Wine Enthusiast Guide* on a Pocket PC

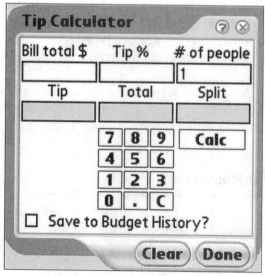

Illustration 11-2 DinersChoice's tip calculator on a Palm Powered device

Good food also deserves a good tip, despite Baedeker's protestations. Both Pocket PC and Palm Powered handheld computers include a calculator. To calculate a 15% tip, multiply the total bill by 1.15. In Europe, the tip is usually 10%, and this can be calculated by multiplying by 1.1. Things become more complicated when dividing the bill among the different diners, which is why you may prefer DinersChoice (www.l3solutions.com).

Cooking at Home

Although those in the food industry may need to pay the full $19.95 for the latest version of MasterCook Deluxe (www.valusoft.com), older versions may be all that home chefs require. They cost just $9.95 and allow you to create a custom shopping list on your handheld computer for the recipe that you want for your guests. Finally, for the precision barbecuer, BarBQ buddy (www.mobiledynamo.com) has great tools.

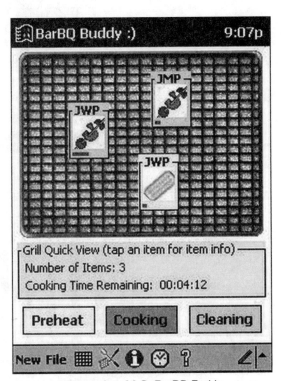

Illustration 11-3 BarBQ Buddy

Sadly, food comes with calories, and at some point we have to count them. The U.S. Department of Agriculture makes National Nutrient Database freely available[1] for Palm Powered devices.

The software is not available for Pocket PCs, but it is better to get the full Diet & Exercise Assistant software (www.keyoe.com) anyway because a healthy lifestyle requires exercise to go with the balanced diet. The software is available for $19.95 for Palm Powered and Pocket PC devices, and a PC version costs another $9.95. This means that you can conveniently log what you eat and how much you exercise using your handheld computer. The software's historical trends work well on

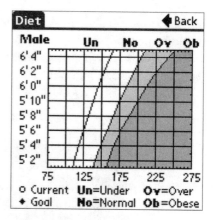

Illustration 11-4 Diet & Exercise Assistant on a Palm Powered device

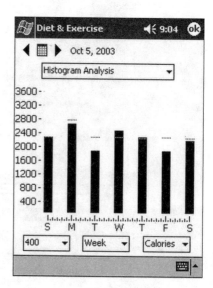

Illustration 11-5 Diet & Exercise Assistant on Pocket PC device

Illustration 11-6 Handheld diet diary

the devices, but the PC version is even better for this because of the large screen.

Of course, as Baedeker would tell you, these are not the only choices. Pocket PC Central lists a nice selection for Pocket PCs,[2] and it even provides another list for Palm Powered devices.[3] Go on—eat and be merry.

FURTHER INFORMATION

- Diet & Exercise Assistant—www.keyoe.com
- DinersChoice—www.l3solutions.com
- MasterCook Deluxe—www.valusoft.com
- USDA National Nutrient Database –www.nal.usda.gov/fnic/foodcomp/srch/search.htm
- Vindigo—www.vindigo.com
- Wine Enthusiast Guide—www.landware.com
- Zagat—www.zagat.com
- Zagat to Go—www.handmark.com

END NOTES

1. www.nal.usda.gov/fnic/foodcomp/srch/search.htm
2. www.pocketpccentral.net/software/diet.htm
3. pocketpccentral.net/palm/software/diet.htm

PART 2

TWO HANDHELDS ARE BETTER THAN ONE

Chapter **12**

Case 1—Menu Entry

Patrick Leon Esquerré says he was just a "French country boy" when he arrived in the USA. In Texas he met and learned from Stanley Marcus of Neiman Marcus fame. Although he had never been a baker, a banker, or a restaurateur, he did have his upbringing in the Loire Valley region of post–World War II France for inspiration as he opened his first restaurant in Dallas in 1982. Now, with 60-plus locations in five states, la Madeleine chain is thriving and handheld computers are a key part of its next wave of innovation.

Their partner in this is MICROS Systems, Inc., a leading provider of information technology solutions for the hospitality and retail industries. Over 220,000 MICROS systems are installed in restaurants, hotels, motels casinos, and retail foodservice operations in more than 130 countries and on all seven continents. A British company had already had success with its solution for menu entry handheld computers and kitchen display systems (KDS), and la Madeleine is one of the first major chains to use MICROS handheld systems in the United States.

The Vision

Mohammad is a big fan and a regular customer of la Madeleine. In his early visits he enjoyed the relaxed simplicity of dining. Close to the entrance he and other customers would order different hot food while adding salads and other ready-made items to their tray. With the hot food order the staff would give him a wooden block with a letter to identify his order and a paper receipt with the cost. At the end of the line he would pay the cashier for all the items. Wait staff would later bring the food to his table, identifying him by the wooden block on his tray. Mohammad introduced Susan to la Madeleine when her travels to meet him face to face for the first time coincided with her first MICROS user conference in Maryland.

At the conference, Susan with MICROS's gracious support, met two potential subjects for the menu entry case. One was a company from England that had had major success with menu entry handhelds and kitchen display systems (KDS), and one was a national group from the United States. Susan gasped; an image flashed before her eyes—this national group was the one Mohammad had introduced her to this past weekend, la Madeleine. Mohammad's favorite place had now become a potential case participant. To fuel this connection, when Susan visited San Antonio one month later for the American Culinary Federation Accreditation Commission meeting, she stayed at a bed and breakfast and found out that the hostess's preferred spot to eat was—yes, la Madeleine.

While writing this book Mohammad noticed the changes around him. On entry, a staff member would take his order on a handheld computer. He still gets a wooden block, but by the time he arrives at the cashier, the details of his orders have already arrived from the handheld computer, and the kitchen staff also have the same details.

Mohammad still enjoys the same experience, but the restaurant has greater efficiency, and the head office has far more information in its computer systems for analysis.

The Team

As you might expect, it takes a lot of people to make such an ambitious project a success. Susan's first contact was with Team MICROS, through Irwin Fein, a Long Island representative, who highlighted the chains putting the solution to use. Irwin led Susan to the MICROS Major Account Restaurant Group. This group was critical to Susan's understanding when she had to baby-sit a handheld and two batteries for testing on a long drive back to New York and provided her with support by sharing MICROS screen shots and information.

George Popson is senior director of information technology at la Madeleine. George's first session with Susan was informative, honest, and refreshing. He immediately brought in from la Madeleine Richard Hodges, director of learning, a consummate professional and a critical link to Susan's further understanding of the application of this technology in la Madeleine's "world."

The Challenge

At la Madeleine communicating orders from hostess (order taker) to food preparation area was disconcerting due to a unique physical plant design and might have been contributing to missed sales and increased ingredient costs. The challenge at la Madeleine was to train those who had used handwritten order pads to embrace handhelds. Foodservice establishments desire to overcome such challenges unique to their situations: poor customer service, decreased productivity, and increased costs. These problems (and more) have a solution.

The Solution

Wireless—the use of handhelds may support the already established physical layout, increasing throughput. Although not hard to use, using handhelds takes some practice. But their use, and remote printers, optimized the bakery, café, and bistro's speed of service.

For la Madeleine, the handheld is designed to input orders for items prepared in the sauté, sandwich, and in some locations coffee bar areas. Other foodservice establishments have menu entry requirements that include quick serve, à la carte, self-serve, fast food, and bedside service in hospitals and health-care facilities; handheld capabilities can adapt to any menu and service format, varying slightly, but the principles remain the same.

la Madeleine's first stage of convincing the "financial folks," the technology information personnel, required the success of a strong pilot case. Embracing this concept required the support of the learning director and a location that already had very good systems and food cost controls in place. They wanted to show that the technology would make the most productive store even more productive. A good rationale!

The use of a handheld menu entry order system was first piloted at la Madeleine's lead operation and soon extended to five additional locations with the intent of converting all cafés to handheld capabilities. The desire to replace the hot food orders' duplicate "chits" or receipt paper system (one copy to kitchen and one copy to customer) with remote printers and handhelds was taking root.

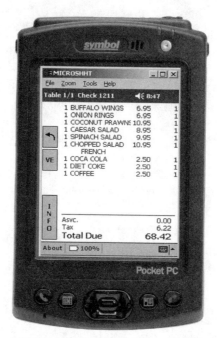

Illustration 12-1 Symbol Pocket PC

Illustration 12-2 MICROS Workstation 4 POS

A. HARDWARE

Handheld unit selection may vary with users. For various reasons, Symbol (see Illustration 12.1) is the primary example of the handheld unit used in this case. This ruggedized unit (more on ruggedized options in Chapter 14, "Case 3—Inventory") will survive multiple drops of up to four feet on concrete, withstanding temperature ranges from 14°F to 122°F. It will survive some direct sunlight, but exposure to prolonged direct sunlight is not recommended.

Additional hardware included MICROS Workstation 4 (see Illustration 12.2) and MICROS Eclipse point-of-sale (POS) device (see Illustration 12.3), which may also include credit card over Internet Protocol (IP) capability, communicating directly with the World Wide Web.

In the food preparation and receipt areas, station printers may include remote printers and thermal printers (see Illustration 12.4 and 12.5). The printers are connected by wireless technology to the handhelds and POS.

This communication required Wireless Access Points situated strategically to receive the handheld orders so that orders could be efficiently received by the cooks, updated with changes, printed for the customer, and processed by the cashier for payment (see Illustration 12-6).

Illustration 12-3 MICROS Eclipse Point of Sale

Illustration 12-4 Station Printer

Spare batteries are required for maintaining "live" handhelds for the ordering process. It is highly recommended that you have backup batteries charged at all times to support all of the features on your handheld (see Illustration 12-7).

Hand Held Cradle Charging Stations (see Illustration 12-8) do just that. They "house" your handhelds for times of nonuse during the day and night and provide a safe, secure, and live environment collecting energy

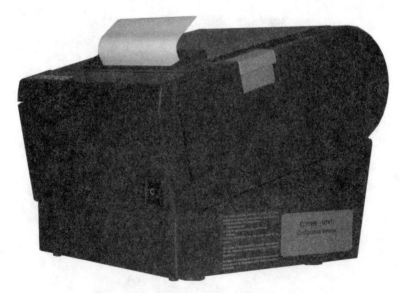

Illustration 12-5 Remote Thermal Printer

Illustration 12-6 Cashier processing payment (Courtesy Corbis Digital Stock)

for continued use. It is advised that charging stations are available in front of the house locations near the greeter station (keep out of sight and secure), additional ordering locations, and a locked environment overnight.

B. SOFTWARE AND REFERENCE MATERIALS

Software includes the Mobile MICROS third-generation handheld system and MICROS 3700 POS. MICROS runs on the Windows Pocket PC Platform and was developed specifically for table service and quick serv-

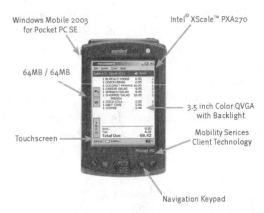

Illustration 12-7 Handheld with identified features

MC50 4-slot
   Cradle ⟶

Illustration 12-8 Handheld Cradle Charging Stations

ice restaurants and includes all the features restaurant managers want in rugged and reliable packaging. The MICROS 3700 POS system features various applications to streamline back-of-the-house operations and provide current, up-to-the-minute reporting analysis. In 2005, la Madeleine upgraded from the Pocket PC2002 operating system to Pocket PC 2003.

An invaluable resource for this case and for users at la Madeleine was the training guide they developed, *HandHeld Resource Guide 2005 La Madeleine de Corps, Inc*. Training guides available in formats that may be utilized by handheld environments will be a useful tool for any new users of such technologies.

C. TRAINING

la Madeleine has a blended strategy for implementation. They incorporated online Webex previews by market, in market training classes, and in on-site training for two days in each location as they went live with the hardware. As Richard Hodges notes, "All elements of the training played a part in the rollout—some elements address the technology phobias, but the on-site training not only helped as a security blanket for those fears, but ultimately was the critical element to the success of the rollout."

Be advised that the information here is a synopsis of the use of one system and in no way captures the entirety of the application's capabilities and flexibility for other foodservice operators.

General . . .

It is very important that all users have the proper job codes in MICROS when using the handhelds, since these affect the screens that are viewed on handhelds. Equally important is the care of the ruggedized handheld. They are rated for occasional dropping and limited exposure to moisture. During normal use, the device should remain with the individual using it (host or user) or in the cradle. Use of the compatible stylus provided with the handheld is required, whereas any pens, pencils, markers, or other sharp pointed materials (no knives please) will damage the screen. Cleaning the handheld screen with a soft cloth diluted with the recommended cleaner and using a damp (not wet) cloth to clean the body of the unit is critical for sanitation and safety reasons. Never submerge the unit in water.

Case Specifics . . .

Note that the screens represented here will vary from foodservice location to location and will be modified to fit the users' specific needs by the technology company providing these services.

1. Hello

The first step is to sign in on the handheld and enter a check ID (see Illustrations 12-9 to 12-11) The handheld processes orders faster and efficiently and is an electronic version of the order pad and the pneumatic tubes of yore, but it has to know who it is, so ID is critical.

2. Key Identification

The main screen for ordering is depicted here (see Illustration 12-12). The arrows allow you to scroll up and down if there is more information than is visible on the screen. The Cancel key allows you to cancel the transaction before any food has been ordered. The Enter key is used when your selection has been made. The Spec Pr key is for special preparation instructions. The Finish key is used when the order is ready to be sent to the sauté area. Pressing the Void key twice allows you to remove the previous item entered. Remember that handheld screens can be modified by MICROS to suit your individual needs.

3. Placing an Order

Okay here we go! They have already entered a check ID for your partner, who ordered a salad, Chicken Caesar, and now you want to order your meal. Let's walk through this specific order together. You're "hungry" for beef—well that in la Madeleine "handheld talk" is: Dinner/Hot Frnch (see Illustration 12-12).

Illustration 12-9 Signing In

Illustration 12-10 Begin New Check

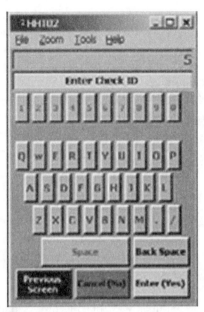

Illustration 12-11 Adding an ID for hostess

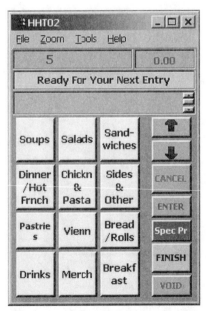

Illustration 12-12 Main Order Entry Screen and Dinner/Hot Frnch

Touch that key, and the next screen will appear in the display (see Illustration 12-13). The key BEEF TENDER is it—but wait, you want to select the beef with red wine demi on the side. The order taker presses the Prep key, and the chosen variation appears on the next screen (see Illustration 12-14), thus telling the sauté cook to modify that entrée. Touch Finish to begin processing this order.

A lettered or numbered 4 × 4 inch block is placed on the customer's tray, which will be utilized by the cashier. This is a carryover from the older system and still serves as a very functional part of the current process, as described later. The blocks also play a part in the experiential process for the guest. They make the transition almost seamless for the guest. Few noticed that they were not handed a receipt, but all would have noticed not receiving a block. la Madeleine's marketing team and operators believe that the "block" system is a part of their heritage—which technology has now enhanced.

4. Time to Pay

Now that you're familiar with placing orders using the handheld, let's pay up. Remember that at la Madeleine's an ID number has been assigned to the block, creating an "open check" in the system. The cashier at the end of the service line monitors the amount of "open

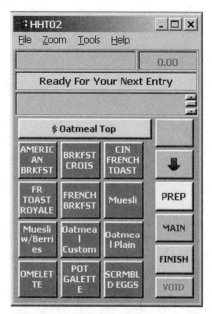

Illustration 12-13 Dinner/Hot Frnch

Illustration 12-14 Beef Tender

checks" in a POS. The guest has already placed an order with the host or hostess upon entering the service area and received a block for retrieval. What is ideal about this is that the check remains "open" until the guest gets to the end of the line. All times are recorded in the system with color-coded cues identifying under/over service times on the checks. Green is "grand," yellow is "hmm, what is going on?," and red, "Let's fix it, now!"

When the guest arrives at the register the block(s) on his or her tray have corresponding squares on the register's screen. The cashier touches the screen, and the "open" check and everything that entered on the handheld—from the first greeting to the last soft suggestive sell—appears. Additionally, foods the customer may have picked up in the tray line after ordering the "made-to-order hand-finished items" can be tabulated and totaled by the cashier (see Illustration 12-15 and 12-16).

D. RESULTS

The first pilot at la Madeleine was successful! Handheld terminals for ordering resulted in increased hospitality, improved speed of physical plant output, improved guest check building, and decreased in-

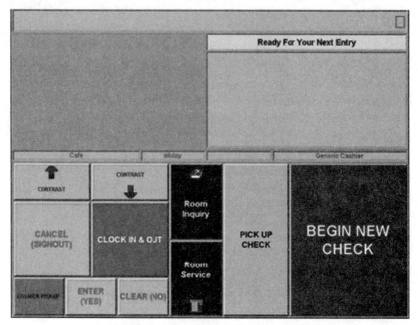

Illustration 12-15 Open Check on POS

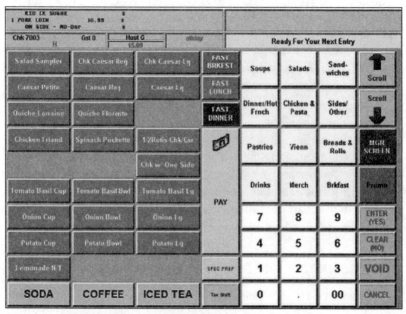

Illustration 12-16 Cashier Pay Screen on POS

gredient costs. The handheld device is a way to process guest orders faster and more efficiently, affecting sales and ingredient costs.

The impact on ingredient costs came from two factors. First was "missing sales" that were now recorded. Cashiers only had to input half of what they used to ring up. Sales and orders were more accurate, because teas, sodas, and side items were not being missed. Sales went up, and ingredient costs went down. Second, ingredient costs from the selling of suggested higher-return items impacted sales. Cashiers had additional time to talk with the guests, and as a result of "suggesting" higher-ticket items, less water and more drinks were sold, as reflected in la Madeleine's menu mix reporting.

Remember that in other types of foodservice establishments, depending on the menu entry requirements to include quick serve, à la carte, self-serve, bedside, or other service, the handheld's capability to adapt to any menu and service format of menu entry may vary, but the principles remain the same. The results are better customer service, increased productivity, improved accuracy of orders for the kitchen with printed, not handwritten, tickets, cost savings, and fewer mistakes on the guests' checks.

As with any new technology, change is inevitable, and as we all know many employees perceive such change with mixed emotions, and justifiably so. But it only took one weekend for employees to learn the system. The use of the technology was intuitive, and it was management that leaped the greatest hurdle. Dealing with separate checks after the order—every server's nightmare—was one of the hardest challenges, but it was accomplished with a few keystrokes. A user was born!

At the time of the writing, fifty-seven of sixty-three locations were on target for a rollout of this technology (all new bakeries will have them). It was projected that a small remaining percentage of low-volume and mall locations would not receive immediate implementation of handheld systems.

Most exciting for upper management is that at the end of every day the information first captured by the handheld in these locations is downloaded from the handheld to the PC server in the bakery. Then all POS information is downloaded from the bakery to their offices—and then sent out by e-mail to the Treos (Palm handheld). They have communicated their performance full circle, to say the least, by using handheld technologies.

la Madeleine projects that after this rollout has been tested and continues to prove financial viability, they will then pilot MICROS kitchen display systems (KDS). Well, that's another book and case study.

Costs

I do want to say that costs for this system were hard to quantify (confidential), but I can share some basic information on what some of these costs were in 2005. If I were going to write a grant for the testing of the use of one handheld, it would look very much as follows. First, you would have to have a compatible personal computer with a Pentium processor at least. Additional costs include: foundation software for MICROS ($1,000), a handheld ruggedized Symbol ($1800), the licensing fee ($350), a Symbol portable printer ($600), two batteries ($75 each/$150 for two), a charger ($160 to 200), a single-user software database to experiment with the system ($800), and institutional training support per diem ($100 per hour), which totals $4,000 plus. I am sure that there are additional access point hardware and licensing costs, but after I tabulated this, I stepped back, saved my grant for another opportunity (check out Chapter 20 for more on this), and reached out to locations that were piloting/using and voilà!

LESSONS

- Operations personnel and executives must be involved in all decisions.
- Support is critical at all levels from executive boards to operations.
- Learning curves must be acceptable.
- Understand that there will be resistance to technology (not just change), and make a plan that will address it.
- There are different skill levels (especially with a multigenerational workforce)—training should address all levels. Many times, the associates will take to it before the managers—training should be specific to each group.
- Efficiency is not always an inexpensive/expensive endeavor.
- Team building takes time. Synergy results.
- Do your homework, and check out the competition and their service records.
- Smile when talking about technology, and remember it is only as valuable as the user who is valued.

5 QUESTIONS FOR STUDY

1. What additional training materials and resources might you develop for visually and hearing challenged employees and customers who are using menu entry handhelds?

2. Who are all the "players" when a decision to use this technology comes from the information technology/systems department?

3. Where might one find additional companies that are selling and using this technology? Who are these companies?

4. When is it an appropriate time to embrace a reticent ten-year employee to make this transition from dupes, chits, or paper orders to the use of handheld menu entry?

5. Why is this form of technology (menu entry with handhelds) critical to the success of foodservice operations today? What do you think is the future for handhelds?

And More Thanks

In closing, I would like to mention other technology solution providers that have embraced and shared additional information on menu entry handheld applications and capabilities and further acknowledgments.

Software solution providers include: CBORD Nutrition Service Suite: Bedside Menu Entry (and more), Cookenpro Commercial Suite (Cookenpro and a PDA application—Barrington Software), ChefTec PDA physical recipe writing and viewing application, Digital Dining, Ameranth Wireless 21st Century Restaurant Software, and ALOHA—Radiant Systems, Inc.

Hardware solution providers include: ruggedized options—Symbol 884, SPT 1800 or 1846 or nonruggedized COMPAQ IPAQ, Dell AXIM X50 or higher, Aceeca Meazura, Sony, and Palm Tungsten T-5 OS 3.2 or higher.

FURTHER INFORMATION

- Aceeca—www.aceeca.com
- Aloha-Radiant Systems Inc.—
  www.alohapos.com/public_site/about.htm

- Ameranth—www.ameranth.com/home
- Barrington Software Inc.—www.cooken.com
- Cheftec PDA—www.culinarysoftware.com
- Compaq—www.compaq.com
- Dell—www.dell.com
- Digital Dining—www.menusoft.com
- la Madeleine—www.lamadeleine.com
- MICROS—www.micros.com
- Palm—www.palm.com
- Sony—www.sony.com
- Symbol Technologies—www.symbol.com
- The CBORD Group Incorporated—www.cbord.com

ADDITIONAL ACKNOWLEDGMENTS

MICROS TEAM

Louise Cassamento, vice president of marketing

Peter Rogers, vice president of investor relations & business development

Alan Hayman, executive vice president of restaurant sales and solutions

Chapter **13**

Case 2—Hazard Analysis Critical Control Points (HACCP)

"You can't recall food in space," said NASA's scientists. So, in the 1960's they worked with the Pillsbury Company to develop a control method for preventing biological, chemical, and physical hazards from entering the human food distribution chain. The results eventually became Hazard Analysis Critical Control Points (HACCP), a thorough and sophisticated process.

HACCP includes seven guiding principles supporting a systematic approach to the identification, evaluation, and control of food safety hazards. HACCP assurance is a rational way to ensure food safety from harvest to consumption—a preventive system to reduce the risk for food-borne illnesses. These guidelines were proposed and adopted in 1997 by the U.S. Food and Drug Administration (FDA), U.S Department of Agriculture (USDA), and the National Advisory Committee on Microbiological Criteria for Foods and based on the 1995 Food Code vm.cfsan.fda.gov/~comm/nacmcfp.html.

When managing HACCPs, one provides records indicating that an individual—a chef or foodservice personnel—in a foodservice operation is monitoring, and capturing temperatures when appropriate, at critical control points: hand washing, holding food, refrigeration temperature data, hot and cold, and the like within safe guidelines. Appropriate food handling, monitoring, and recording reduce the margins of error and potential for food-borne illnesses. This monitoring reduces the possibility of fines and indicates that an organization is food-safe compliant. A record of training and activities is time stamped, and corrective actions are noted and stored in a database proving that an HACCP plan and system are in place.

From 1995 to 2004, the HACCP platform changed, and by 2004 foodservice facilities regrouped. For example in 2005 public schools food programs were required to implement an HACCP-based food safety program (http://sop.nfsmi.org/HACCPBasedSOPs.php).

Still, even by 2005 few in the foodservice industry embraced HACCP implementation on a voluntary basis, mainly because it is so expensive to do.

The Vision

Sodexho is one of these few. It is not mandatory for them to monitor their critical control points (CCP's): hand washing, cooking, hot and cold holding, cooling of foods, reheating, and refrigeration. But Sodexho wanted to do this as part of its own quality consciousness. The need to meet changes in the food code, consolidate paperwork, and focus on the CCPs designated by Sodexho sparked an interest in mobile data collection as a tool.

The Team

The players for this case comprised three major teams: the leading food and facilities management services company in North America, Sodexho; a proprietary software and services provider, NextStep Edge Technologies, Inc; and a Sodexho foodservice client, ElimPark Place. Sodexho, a member of Sodexho Alliance, provides foodservice offerings for campus dining, school and corporate dining, concessions, food and nutrition services, and much more. I found them by surfing the Internet and "googling" handhelds and food safety.

In Washington, DC, March 2005, the Food Safety Summit website identified a team of three presenters sharing information on mobile data collection and monitoring for HACCP. John Zimmermann, Sodexho zone director, QA & food safety, was listed as a presenter for the Food Safety Summit, so Susan e-mailed him to get the ball rolling. She had worked for Marriott in Washington, DC, in the Business and Industry division at the World Bank, Treasury and Monetary Fund, and felt we shared a common thread, since Sodexho had merged with this Marriott division. John was kind enough to connect her with his colleague, Mike Dunn, Sodexho director of product quality assurance, who had actually been the voice at the summit presenting "HACCP in Your Hand: Mobile Data Collection and Monitoring."

Without Mike, this case might not exist. His forbearance with the crazy chef's (me) e-mails and his gracious follow-up with NextStep and ElimPark Place have earned him a place in my list of doers! Mike also shared the e-mail addresses of two of the other presenters, one that I contacted and will mention later in the "and more thanks" section, Kimberly Eifler, director of commodity quality, Darden Restaurants. Shaji George, food safety and health technical manager, Walt Disney World, was kept in the wings in case I needed additional support. Mike noted that they were great presenters with more information, so I had a soft place to fall.

Mike is a former USDA inspector and worked for Quaker Oats—some resume! We scheduled an interview after I forwarded him a series of questions:

1. What hardware are you using?
2. What software?
3. Typical use in day-to-day operations. Describe.
4. Audience using handhelds. Describe.
5. Bilingual capabilities? Other learning abilities?
6. Are your purveyors HACCP approved?
7. Handheld software monitoring temperature only or time and temperature?
8. How long did you test/pilot hardware and software?
9. How many foodservice sites are utilizing?
10. Wonderful things about using this hardware/software.
11. "Areas of opportunity" (weaknesses) for hardware/software.

Once we connected, the nuggets flowed forth with the remaining team members falling into place. Mike introduced me to NextStep Edge Technologies, Inc. Sodexho selected NextStep because it was easy to use, cost-effective, had a palm OS platform, and was flexible. NextStep's technology provides a suite of proprietary products that work on virtually any Pocket PC and Windows NT– and XP–compatible system. They provide services that are web based and handheld compatible, allowing the capture of aggregate data in close to real time. Susan contacted Joel Spada, vice president of operations, by e-mail, and he directed me to Gerard Colucci, founder and visionary.

What a busy individual, always on the road—we talked briefly but "many times." He mentioned that Sensitech (www.sensitech.com) was one of the pioneers embracing HACCP and the base or model, for NextStep Technologies, Inc. The remaining participants made the hand-held come alive with my visit to Sodexho's client, ElimPark Place, a not-for-profit, interdenominational, nationally accredited continuing care retirement community in Connecticut, to see HACCP on hand-helds at work.

Joseph Cuticelli, recently promoted from ElimPark to area direc-tor for Sodexho, was Susan's initial contact. He connected her with Chef Adam Hart, newly positioned at Elim (six weeks). But it was Sous Chef Jesus Medina III who was her main guide. Thank goodness for all the support of sous chefs in the world, the backbone of foodservice kitchens. But more on that as we meet the challenge, identify the so-lutions, and more.

The Challenge

Measuring, monitoring, and logging internal food temperature data compliantly with state and federal HACCP reporting requirements on paper proves to be an overwhelming task. Food temperature con-trols, refrigerator and freezer temperatures, dish room and sanitizing checks, and monthly audit checks are just a few of the individual ac-tions necessary to operate a food-safe facility. Multiply that times thou-sands of facilities and you have a paper collection nightmare.

The Solution

Capturing data on handhelds to measure, monitor, and log internal food temperatures compliantly with state and federal HACCP reporting requirements and using web-based real-time retrieval at corporate head-quarters is an answer. The tools necessary to complete these tasks in-clude web access, hardware, software, training, and support.

A. HARDWARE

The Dell Pocket PC Axim 30 (see Illustration 13.1) replaced an HP iPAQ replacing an even earlier Palm OS handheld. Using the Palm for this application, as Mike suggested, was like using DOS rather than Windows. A standard temperature probe from Tangent Systems Inc., Versid DT-10 is cost-effective and uses a Windows Mobile OS platform[1] or "K" type probe, as mentioned by Mike.[2] He advised that lighter probes may be fooled, so do your homework.

A wireless Bluetooth probe is in the developmental stages and may be available and affordable by the time that this book is in your hands. Palm OS users may prefer an affordable Themapalm probe.[3]

Remember, use a temperature device that is recommended for your handheld software. Handheld cradles, charging units, or charging plugs that come with Palm or PC Platform handhelds are standard purchases and have been covered in previous chapters.

B. SOFTWARE AND REFERENCE MATERIALS

Materials referenced include Mike Dunn's Food Safety Summit PowerPoint presentation, *The Culinary Data Assistant (CDA) Manual*, and information captured at ElimPark.

Sodexho's rollout in 2003 used the CDA in a Palm OS. With over 200 users, replacing paper food temperature logs for cooking, holding, cooling, and reheating presented an interesting trial. Advantages included paperless temperature data, making it more fun and allowing

Illustration 13-1 Dell Axim

for more data collection. However, since it was not web based and not as user-friendly as it could have been, the mobile collection entered phase two.

In August 2004, the CDA was customized by NextStep Technologies, Inc. to include Pocket PC–Windows Mobile and tested with a handful-plus of Sodexho locations. A friendly web-based program replaced the software on the computers, allowing Sodexho to fully appreciate the true mobility of a handheld device.

NextStep Technologies, Inc. software HACCP Data Assistant (HDA) was modified for Sodexho and included a temperature acquisition or HACCP log for handhelds. Additional web-based services include a quality management systems (QMS): forty quality assurance food-safety-related forms used for tracking additional inspections. Sodexho embraced these forms to track inspections, enhancing the handheld's capabilities. The mobile handheld communication with a web-based location burgeoned into an open window—an administrative website accessible to Sodexho headquarters quality assurance staff. Operator data was easily accessible on a daily basis and logged and collected in a database.

NextStep kindly provided us with the CDA manual and access to this online environment so that Susan might test synching to the web-based tool with my personal Dell Axim. But seeing it in action at ElimPark Place was truly memorable.

C. TRAINING

As Susan entered Chef Jesus Medina III's kitchen at ElimPark, the first thing she saw was a moving ticker-tape-type sign that contained various location temperatures and colors—Dining Room 38.8°F (cold for a room, she thought). But no, it wasn't the temperature of the room. but the refrigerator in the dining room. All the refrigeration units were connected to this sign via a computer. Green was "okay", yellow, "let's monitor", and red, "immediate care!" Memories, these were the universal colors used for communication in Chapter 12, Case 1, menu entry at la Madeleine.

Chef Medina was preparing breakfast—pan savers, alcohol swabs, and a handheld, not your usual breakfast condiments. He finished his production and graciously took me under his wing and began to instruct me on how to use this device.

First, you connect with the web-based operator's website, entering a user name and password (see Illustration 13.2). The online web tool

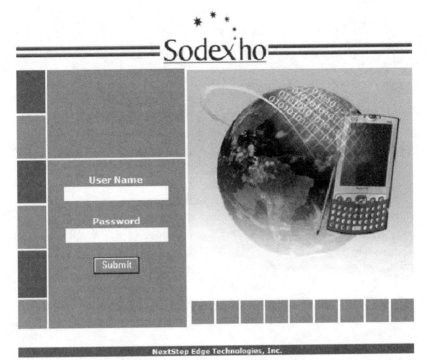

Illustration 13-2 Operators website

is where you manage your data for use on your handheld, which includes maintaining CCPs, HACCP, menu plan groups, food items, corrective actions users, menus, week-at-a-glance information, and form creation such as temperature logs and the like.

After you sign in and click Submit, you will be taken to the page where you can maintain CCPs (see Illustration 13.3) and HACCPs

| CCP's | HACCP Categories | Menu Plan Group | Food Items | Corrective Actions | Users | Approve Items |

Critical Control Points

Critical Control Point		
COOKING	Edit Delete	
HOT HOLDING	Edit Delete	
COLD HOLDING	Edit Delete	
COOLING STEP 1	Edit Delete	
COOLING STEP 2	Edit Delete	
REHEATING	Edit Delete	

Add a Critical Control Point

Critical Control Point Name: [   ]

[Add]

Illustration 13-3 Maintain CCPs

⬛HACCP Categories

HACCP Category					
BEEF, VEAL, LAMB (Roa	Update Cancel Delete	Add Rules	View Items
BEEF, VEAL, LAMB (Steaks & Chops)	Edit	Delete Add Rules	View Items
COLD FOODS (Deli, Salads, Veggies)	Edit	Delete Add Rules	View Items

Illustration 13-4 Maintain HACCP categories

(see Illustration 13.4), including performing additions, deletions, and edits.

Adding an HACCP rule to include temperatures and holding information is managed here by selecting an HACCP category. In this example, Beef, Veal, Lamb (Roasts) was selected (see Illustration 13.5).

In the Menu area, you have two options; either choose No Template, where you can create a new menu from scratch, or click on the drop-down arrow and choose an existing menu you want to copy (see Illustration 13.6). Remember, this is web based, so whoever has the technology and is authorized can make these change anyplace they have a computer and the web!

At ElimPark Place, the user would download the daily "add menu items" information (see Illustration 13.6) onto the handheld as programmed by corporate for that location (or independently if so designated). By selecting the forms tab button on the website, the user can manage forms and/or view the data collected. Once this is done, the operator taps Sync on the handheld to exchange data with the website. Tapping Forms accesses food-safety forms (see Illustration 13.7).

Set HACCP Rules for BEEF, VEAL, LAMB (Roasts)

Critical Control Point	Low	High	Rule		
COOKING	145	999	4 Minutes	Edit	Delete
HOT HOLDING	140	999	140F or higher; check every 2 hours or beginning and end of service	Edit	Delete
COLD HOLDING	0	40	40F or lower; check every 2 hours or beginning and end of service	Edit	Delete
COOLING STEP 1	0	70	Cool hot foods to 70F within 2 hours	Edit	Delete
COOLING STEP 2	0	40	Cool hot foods to 40F or lower within 4 hours	Edit	Delete
REHEATING	165	999	Reheat foods and leftovers to 165F for 15 seconds within 2 hours	Edit	Delete

Add A HACCP Rule

Critical Control Point: [▾]

Low Temp: [ ]

High Temp: [ ]

Rule: [ ]

[Add]

Illustration 13-5 Adding HACCP rule

Menu Name	Start Date	Number of Days					
Cafe Menu	11/08/04	21		Edit	Delete	Week At A Glance	Add Menu Items
Mike's Menu	11/17/04	5		Edit	Delete	Week At A Glance	Add Menu Items
Test Cafeteria Menu	11/08/04	7		Edit	Delete	Week At A Glance	Add Menu Items
Test2 Menu	11/15/05	7		Edit	Delete	Week At A Glance	Add Menu Items

Corporate Menu Templates

Add A New Menu

Use Current Menu as Template: -- No Template --

Menu Name: Cafe Menu

Start Date: 11/08/04

# of Days: 21

Add

Illustration 13-6 Menu maintenance

Next the operator taps on a form to start data collection. As you can see here, your options include: logs for dish machines, freezers, refrigerators, and sanitizers, and a very comprehensive monthly food safety audit. Tap Home (see Illustration 13.8), and you go to the home page (see Illustration 13.7).

Tapping on menus on the screen shown in Illustration 13.7 brings you to the screen shown in Illustration 13.9. Tap on any menu, and select meal periods Breakfast, Lunch, or Dinner.

Illustration 13-7 CDA Sync and Forms

Illustration 13-8 Logs and audit

Illustration 13-9 Menus and Meal Periods

Tap on a food item, for example Big Apple Burger (see Illustration 13.10).

This will take you to the CCPs screen. You would tap on a CCP if this were the control point you wanted to monitor at that time (e.g., Cooking), and temperature acquisition would begin instantly (see Illustration 13.11).

If the item says PASS, which it did for Yankee Pot Roast (refer to Illustrations 13.12 and 13.13), you repeat the menu item process again. Tap on "check mark" to record data, then Food Items to go back (see Illustration 13.12). Remember, clean your probe with an alcohol swab after each temperature reading. At ElimPark, they utilized three probes: one each for beef, pork, and chicken.

If the temperature is not correct FAILED will appear. The user must tap "X" to select a corrective action (see Illustration 13.13).

The user selects the appropriate corrective action (see Illustration 13.14) and automatically goes back to Food Items. Remember, this is all being recorded for your uploading back to the web. You cannot trick the system.

With all the data collected for the day or an agreed-upon time or meal period, the user or identified staff member will go back to the home screen on the PC (see Illustration 13.7) and exchange data

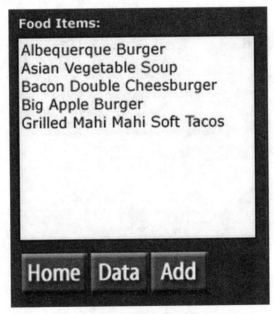

Illustration 13-10 Food Items

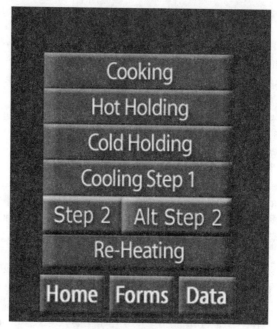

Illustration 13-11 Cooking

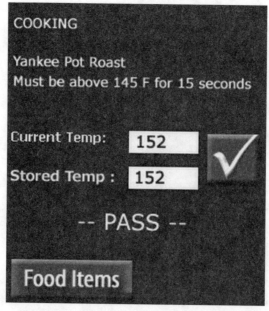

Illustration 13-12 Acquisition and check mark

Illustration 13-13 Failure

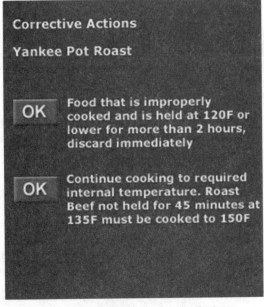

Illustration 13-14 Corrective Actions

(sync) the handheld with the website. The user reenters the user name and password on the website (see Illustration 13.1), selects the Temp Logs tab (not pictured), enters search criteria, and selects View and Print to view the updated document, saving and/or printing a PDF file. Remember that not only do you have this new data recorded but also the headquarters quality assurance staff, district manager, and executive chef, all those authorized, may retrieve the information immediately after you have updated the web-based online files.

Susan asked Chef Jesus what he thought at first and did he mind my sharing this with you. He said he would do most anything requested at work but was a tad hesitant with this new technology, yet at the end of the second day "I wanted to marry the thing!" One day the handheld "crashed," and for two weeks he had to record information by hand, on paper. He kept reminding others to get the device back!

She also met with Chef Adam Hart later that day, who had recently joined the team at ElimPark. He furthered her discoveries on the web-based capabilities once the temperature data was downloaded. He noted that the technology was hard to fool, confuse, or trick, and the website was intuitive and easy to manage. Actually, it only took me a few minutes to find and look at all the reports. His main concern with handhelds was that they may be fragile in a busy kitchen (aren't they all?!) but care and expeditious data collection comes to those who embrace a safe food supply.

Sodexho's solution to this potential increased expense was not to use a ruggedized Symbol, and in Chapter 14, Case 3, I will mention a possible alternative protective cover. Using Dell Axims (inexpensive) rather than the Symbol (expensive) allowed them to purchase more units (about 3:1, respectively), allowing foodservice to cover additional areas in ElimPark. Executive Chef Adam asked the same question that many others chefs did as I was researching this book: when will we get an inventory package on handhelds? Welcome to Chapter 14, Case 3.

Yes, the online website capabilities continued to impress. This went beyond recording temperatures for menu items and included a floor plan of the facility, constantly monitoring temperatures in hot and cold holding areas, including all refrigeration and hot and cold foodservice areas and the dining room. Additionally, service and maintenance logs generated minute-by-minute reports (if needed), and that first cue from the sign that greeted me when I had walked in early that morning now came full circle visually for all to access. (Fresh Loc, www.freshloc.com).

Joseph Cuticelli, her initial contact at Elim Park, had been promoted to Sodexho Senior Services. His presence was felt via wireless, even though he was not physically on-site, when he was notified via his handheld (Blackberry) with a text message that a temperature had entered the danger zone for more than the predetermined acceptable amount of time.

Handhelds synchronized with the Sodexho website, maintaining weekly, monthly, and yearly audit reports and inspection forms, fodder for more knowledge sharing. Sodexho has 8,000 plus menus, 17 HACCP categories, and numerous accounts reporting audits; think of the data retrieval captured by a "tap" on a handheld. All reports are available electronically, and the long-term tracking of trends and "areas of opportunity" (weaknesses) in the system may be monitored in close to real time. But, more importantly, the company has a record of time, temperature, and location for third-party auditing programs such as that of the National Sanitation Foundation (NSF) with whom they must comply yearly.

Future wireless data collection with Bluetooth may require a different handheld device, though most Palm and PC operating systems embrace Bluetooth today. The future for HACCP on handhelds will be real time once the technology catches up and is affordable for foodservice operators.

We work with such small margins and yet are responsible for a safe food supply as we serve at the end of a long and multifaceted food chain. You take the temperature in your foodservice location, and corporate or the manager knows seconds later that you are doing a great job! And the world is a safer place due to your diligence.

P.S.: yes, roast beef rounds take more than two hours to cool down in refrigeration, so HACCP freaks out every time (more than two-hour cooling rule?). Maybe there is a blast chiller in your future!

And what does the future hold? Sodexho and NextStep look forward to wireless data collection (real-time Bluetooth), menu imports to websites, and trend analysis to name a few features. NextStep is testing an Excel capturing system for inventory and ordering on handhelds, and, who knows, this may somehow be integrated. This was not available for the Chapter 14 case, yet I was fortunate to cross paths with others (CBORD) that met our needs.

LESSONS

- Removes mountains of paperwork for managers.
- Logical and fun hardware, software, and training meet generational needs.
- Customers and clients expect growth with technology in a technological world, so this is a good sales tool.
- Foodservice, as an industry, is catching up to the greater technological world, and professionalization comes with acceptance and practice.
- Return on investment with low margins is critical to understand for successful foodservice accounts.
- Multiple checks and balances using technology secure your standing with your accounts, especially populations that may be older, ill, or physically compromised in some way.
- The business world is driving technology—foodservice must embrace it. (Yes, I said this again but in different words.)
- Costs are constantly changing, so keep current, know the competition, and begging sometimes works!
- Gerard from NextStep: HACCP monitoring in the past cost $27 a day; now it is $2 plus pennies.

5 QUESTIONS FOR STUDY

1. How do you convince chefs that sanitation and safety are priorities, as important as the appearance and taste of food? What do time and temperature have to do with food safety and is HACCP a law?
2. Who besides your immediate supervisor may be impressed with your handheld use, especially if you get a passing score?
3. What might be helpful if you discover that a customer of your restaurant claims that they contracted food poisoning from your facility because they ate your cold chicken soup last night?
4. Where is a Bluetooth temperature probe available for purchase? Is it available yet? Would a cold chain company utilize it first?
5. When you have finished reading this, go into the Internet and see how many handhelds are now available for HACCP monitoring. Name them. How much do they cost? Do proprietary companies

still control most of the solutions or can you buy these tools "off the shelf"?

And More Thanks

Kimberly Eifler, director of commodity quality at Darden Restaurants, was Susan's first contact for HACCP handheld information, and although we did not select that company for this case, her information was invaluable. They had used a HACCP system for four years. At Darden on the supplier side, they are required to have a HACCP program. As a result of using central distribution centers and direct shipments, supplier control of time and temperature processes are easily monitored. Random (sample) temperature evaluations by the home office are done and studied, supporting the complete control of the process. Darden has no franchises, so the chain is not broken. Purveyor samples are also continuously audited.

Their restaurant HACCP handheld environment is bilingual and uses pictures. For example, an image of a chicken appears and temperatures at three stages—pre-prep, prep, and service—are monitored. It is simple in design. Red or green image codes appear: red indicates danger; green is okay. This quick pass/fail method removes language barriers. As many of us can appreciate, this is something quite helpful for our industry. The software utilized is SENSITECH, a cold chain company.

Kimberly supported others' claims of ease and mobility. The one "area of opportunity" she referenced was that Darden's handheld use in restaurants did not monitor both time and temperature. She did not have the hardware device's brand name, but a little research on SENSITECH's website identified a tool called TempTale Manager Software that monitors time and temperature from a TempTale monitor. The monitor uses the Windows operating system. SENSITECH tools appear to be proprietary.

In closing, it may be helpful to reiterate that Susan's method for first communicating with all the people in this case started out with an Internet search in Google. Finding within a few minutes the Food Safety Summit and the handheld HACCP presentation, she sent e-mails out to the three companies mentioned, to either their career or public relations people. It took less than a week for Darden and Sodexho to respond! Thank you again for technology and the users that embrace these tools and methods. To get this information from them to the publication stage– on paper—the creative process took many more months. Is it something about paper?

Other technology solution providers that have embraced and shared additional information on HACCP entry handheld applications and capabilities are included in the further information links.

FURTHER INFORMATION

- Darden—www.darden.com
- Dell—www.dell.com
- ElimPark Place—www.elimpark.com
- Food Safety Summit—www.foodsafetysummit.com
- HACCP-manage.net—www.shecentral.net/haccp/info_handheldspecs-haccp.htm
- NextStep Edge Technologies—www.inextstep.com
- Palm—www.palm.com
- Sodexho—www.sodexhousa.com
- Symbol Technologies—www.symbol.com
- The CBORD Group Incorporated—www.cbord.com

END NOTES

1. www.versid.com/cgi-bin/site.cgi/versid_modules/dt-10s/index.html
2. www.topac.com/thermcontact.htm
3. www.thermoworks.com/products/logger/thermapalm.html

Chapter 14

Case 3—Inventory and Ordering

Have you ever ordered food for 10,000 meals a day, served in 6 locations, 24/7, with over 3,000 food items?

In the late 1980s at the University of Rhode Island, Susan served as production coordinator, working closely with the information technology personnel and the management team to computerize menu forecasting, food inventory, and ordering. Additionally, balancing a master's in computer education and writing a thesis at Johnson and Wales University identifying computerized forecasting technology as a cost-savings tool was cutting edge. This unique opportunity provided an environment where practical application and academic research supported the concept that computerization results in cost savings.

It is now 2005, and the company providing the technology to make things happen in the 1980s is still cutting edge and unique in having practiced, on a large scale, computerized inventory and ordering for more than a decade, creating even newer opportunities for more cost savings.

The Vision

Tentative commitments from three other groups did not materialize. The CBORD Group, Inc. was, and continues to be, this visionary. Susan had met Thomas B. Hilton, CBORD national sales manager, at the MICROS user conference when trolling for information on handheld bedside menu entry. CBORD knew chefs were not benefiting from computerized technological organizational tools. In busy kitchens, a chef on the fly requires easy access to inventory and ordering of foodstuffs. If a food item is not ordered, it cannot be prepared or monitored for safety, and it will definitely not make it to the menu/customer if it's not in the "house"! So, Susan sent an e-mail, and the connection was made.

The Team

Users of the health-care and campus dining Handheld Inventory Management System (HIMS) included: Della Rieley, M.S. R.D., director nutrition services, Saint Mary's Regional Medical Center in Reno, Nevada, and Merelene Stanley, coordinator for dining services, Technology Housing and Dining Service, University at Boulder, Boulder, Colorado.

The technology solution provider was CBORD Group, Inc., the world's largest supplier of food and nutrition software solutions, campuswide ID card programs, cashless dining, and housing management systems (which celebrated their 30th anniversary in 2005). Tom Hilton connected Susan with Rich Higgins, vice president of sales, and Rich H. connected me to Rich Barnes, northeast account representative. She publicly apologizes to both of them for early onset confusion (intellectual interludes, not senior moments). Two, too many Richs made for "poor" distinction on my part. Rich H. provided her with a financial quote for an educational grant (more about grants in Chapter 20) that she was processing for approval to purchase CBORD products for an educational setting. Rich B. was her "eYoda," handheld mentor extraordinaire.

The Challenge

Using inconvenient tools (paper and pencil) for recording inventory and having unprotected access for order management may result in inaccurate inventories and potential security risks. Security risks, too, are an essential consideration when looking at financial data, but this case does not address this specifically. It is an accepted practice when you share data that security is a priority for your solution provider. For this case we were most interested in the direct discussion of inconvenience and inaccuracies in the inventory model for chefs and how to improve this.

Della Rieley, in a phone interview, reported that their challenge was to take printed inventory sheets, utilize them for manual inventory, and enter this data into a computer-based Excel or tabulation program. It was a constant trial. The more numeric entries by "humans," the more opportunity for mistakes, and the more time it took to capture the information, the greater the costs for the organization. Merelene Stanley confirmed these same challenges and also identified that decreased speed and inaccurate calculations were their bugaboos.

The Solution

A mobile tool to include barcoding and handheld scanning devices used with inventory and ordering software increase efficiency and productivity and reduce errors for foodservice operators.

Della Rieley reported that the solution, the use of a handheld, reduced the number of steps in the inventory process. The inventory was taken on the handheld and then synched to the PC (personal computer). The software on the PC immediately calculated the entire inventory once the download was performed. But how?

A. HARDWARE

HIMS requires a handheld/PDA (personal digital assistant) Palm Tungsten E2 or TX model. Specifications in 2005 preferred Palm operating system (OS), version 3.2 or higher, 33 MHz processor, and 8 MB RAM minimum (16 MB recommended). Your cradle selection must be suitable for the operating system of the workstation that will be used. For example, note that Windows 95 and Windows NT do not support USB connections. A barcode reader and USB port are recommended but not required.

The Symbol line of wireless barcode scanners (SPT 1800 Series) are ruggedized computers combining scanning, wireless connectivity, and powerful processing capabilities with Palm OS platforms. Symbols are more expensive, but they have a larger screen and are more robust.

There are many accessories required for complete functionality: a single slot cradle with power supplies, a serial interface cable and line cords, communication cables, USB cables, battery chargers, and a metal stylus. It is recommended that you have extra charged batteries and multiples of recommended styluses for your handheld available in locations where they are easy to retrieve.

Della Rieley uses a Palm Tungsten T2s (without scanner). Merelene happily utilizes a Handspring VisorPro® with a Palm operating system. Scanning was not instituted yet at either location, but Boulder was ready, since VisorPro has scanning capabilities (available using a plug-

in module offered by Symbol). Both locations are interested in insti-
tuting barcoding in the very near future.

Optional items include magstrip readers and pistol grip handles.
Susan recommends keeping your hardware supplier's website in your
Internet Explorer Favorites. You will find many accessories available to
make the handheld more efficient and easier to use. If your handheld
is not ruggedized, getting a protective case for your handheld is recommended. Palm PDA and Sony are not ruggedized, so it is strongly
recommended that you look at the accessories your handheld's website may offer to protect your device.

Recently, we were approached by Zachary Borger of Otter products, a company offering hard case options. The Otter Box 1900 (see
Illustration 14.1) is water-, drop-, dust-, dirt-, and scratchproof. It has
a replaceable screen membrane and a rugged IP67 and MILSPEC 810
rating. It is expandable for use with a CF/SD Bar Code Scanner and
costs $99.95 + $39.95 for scanner pod kit (MSRP, retail). It fits most

Illustration 14-1 Otter
Box 1900 (PDA inside
box is iPAQ2, not com-
patible with CBORD
HIMS)

PDAs with various access points to standard PDA connections and headphone access. An unconditional lifetime guarantee is reassuring.

The Otter Box 2600 is similar to the 1900 in that it is waterproof, airtight, and crushproof. It includes IR beam data and retailed at $49.95 in 2005; it is easily cleanable and may save you thousands of dollars by reducing the need for replacements due to breakage. Success stories include great savings for United Parcel Service, where UPS field technicians were breaking an average of three PDAs a day.

Although these can provide considerable cost savings when buying ruggedized units ($1500 to $2200), Susan did find, after flipping up the screen cover, that the fully flexible membrane on both units was hard to penetrate (tap) with her standard stylus. The best overall application of this hardware will be with barcode scanning. The fact that the Otter can be adapted with a Socket (barcode scanner) that costs $200 to 250 will ease the pain of having to press hard on the membrane to enter data on the screen.

We do look forward to the day that all foodservice items have barcodes, so we can maximize the value of this rugged shell and scanner. This hardware will be especially helpful in freezers (truly) and taking inventory in those lobster tanks while we are falling off the roof of a building! We are just kidding, but not totally. Zachary shared one story of a handheld in an Otter case that survived a fall from a four-story building.

B. SOFTWARE AND REFERENCE MATERIALS

HIMS is a supplemental application to the CBORD's inventory module Foodservice Suite (FSS). This suite includes an inventory management system, a handheld inventory management system, and an HIMS unit software. The program enables recording of inventory counts directly onto handhelds with Palm operating systems. The user downloads an inventory tally sheet onto the handheld, covering storage area food items, and records physical counts; the computer returns data that is uploaded into FSS. This software solution allows you to take a physical inventory using Palm OS technology. Non-CBORD software and equipment includes a Sybase ASA with an Ultralite Mobilink and a handheld unit. If you are considering this purchase, it is interesting to note that FSS has HACCP instructions imbedded into their production recipes.

Della and Marlene spoke highly of CBORD reference materials that support hardware and software integration and training resources.

C. TRAINING

The Handheld Inventory Management System (HIMS) training materials and additional training resources related to HIMS were provided by CBORD and users, including:

1. HIMS eSeminar on CD and HIMS and print support documentation (version 4.0).

2. Observations by HIMS account users.

3. eSeminar, November 2005, "How to Interface The PDA Software with Geri Menu."

4. Prerecorded session placed in CBORD web page archives so that you can revisit them with a password and ID number (only available to users).

HIMS eSeminar on CD and HIMS Printed Documentation . . .

The hour-long CD contains the entire eSeminar discussions of HIMS functionality, including specifics on setup and inclusion of HIMS with the physical operating sequence.

When you initially turn on the handheld unit, an HIMS icon resides on the screen. A downloaded list from FSS must be preloaded by using the hot sync capabilities when you first sign on with the handheld. First, though, you must load all required Palm applications, Sybase Mobilink server, and Sybase Hot Sync Client, before taking an inventory. The recorded CD eSeminar explains this process verbally and illustrates for the user the activity of sharing applications using Internet Explorer, accessing the C: drive, and installation procedures for HIMS on your desktop and Palm. This process will not have to be repeated after initialization.

In FSS, one either creates an inventory or opens a preexisting inventory after the initial start/creation. A confirm selection message appears, showing the date, unit, and title of inventory. Before you sync, you must return to the C: drive, activating Mobilink batch files, which serve as the conduit for the sync. This file has to be opened each time for sync activity. It is recommended that you place this on your desktop as a shortcut. You sync again to download data, and items are sorted alphabetically by storage location or bin. The download includes up to 35 characters for each food item.

The eSeminar step-by-step simulation with a handheld emulator rounded out the presentation. The emulator helps you understand the standard format for entering the inventory (see Illustration 14.2).

Illustration 14-2 Step-by-step simulator

Three food items appeared with descriptors on the screen, with specifications (name, case, or count, and the like). The location where you entered your numeric count (physical count) using numbers is located at the bottom of the PDA screen or in the graffiti area. The graffiti area on a handheld is where you form numbers with a stylus, which may take some practice. A small square to right of the location where the numeric count is entered is tapped, and a checkmark appears. When this action is completed, the display scrolls down, and you continue your physical count. It was emphasized in the eSeminar to *remember* to tap Enter so that the inventory is recorded.

The capability to find/search by name, universal product code (UPC), or storage area is an excellent resource in an HIMS operating handheld (see Illustration 14.3).

When looking for a specific item on the list, you click the Menu button, and a Go To menu uses a find feature. Under "Searching," if you are using a barcode scanner, you can check only one current storage area and search for a matched scanned UPC, or select all storage area if you want to search in all areas for a match. VisorPro or Symbol scanner capabilities will allow you to read barcodes.

You can control display preferences in the handheld under "Form Display" by checking multiple items, displaying three inventory items at a time, or detail if you want to show one item at a time. This is good

Illustration 14-3 Find/searching functions

for those of us who may be visually challenged. Under "Key/UPC Display," a key name or UPC code may be selected, or "No displays of this kind" may be selected if you want to show more characters of the item names.

"Add buttons" will allow you to add an inventory item to the list. The entire Food Item File (FIDF) is on your handheld. For example, chicken base, which you have been inventorying by the case, is now in your walk-in as "each." On your handheld, you can add to your inventory and download the file when you sync to FSS.

Once you have finished entering physical counts, hot sync to upload to FSS on your PC. Once the information is in FSS, it calculates the changes. The user can open, view, or change the information using the "Enter Counts" option if desired. One excellent report resulting from this sync is the Requirements Worksheet (see Illustration 14.4).

Menu- and inventory-driven purchasing requirements are consolidated by purchase group with on-hand, on-order, and committed figures providing the complete status of each food item. Many other excellent CBORD inventory reports can also be generated. If you download the data using a serial port, it will be slower than using a USB port, and if you have over a couple of thousand food items it may take five minutes.

Requirements Worksheet

Worksheet Date: 12/1/1999

End Date: 12/4/1999

Inventory Center: Main Kitchen Production Unit

Purchase Group

Item Name	Order Qty	Calc Requirement	Purchase Unit	On Hand	On Order	Committed	Par Level	Reorder	Linked
Pork and Pork Products									
Bacon Sliced Lay Out 18/22 ct	2.00	2.00	1/15#	0.00	0.00	2.00			Y
Ham Flat Buffet Smoked	1.00	0.99	1/11#	1.01	0.00	2.00			Y
Ham Fresh B-R-T 11# Avg	0.00	0.00	11#/Ham	43.00	0.00	2.00			Y
Sausage Patty 2 oz	1.00	1.00	80/2OZ	2.00	0.00	1.00	2	1	Y
Poultry and Poultry Products									
Turkey Breast RTC	0.00	0.00	4/12-14# AVG	0.78	0.00	0.50			Y
Turkey Franks 8/1	2.00	1.40	10/1#	0.50	0.00	1.90			Y
Turkey Pulled	2.00	2.00	1/10 #	0.00	0.00	2.00			Y
Fresh Fruits									
Apples Red Delicious	1.00	1.00	125/CS	0.00	0.00	1.00			Y
Banana Green Tip Color	4.00	4.00	1/40#	1.00	0.00	4.00	1	1	Y
Grapes White	2.00	2.00	1/22#	0.00	0.00	2.00			Y
Oranges Fresh	3.00	3.00	1/113CT	0.00	0.00	3.00			Y
Strawberries Fresh	1.00	1.00	12/1PT	0.00	0.00	1.00			Y

Illustration 14-4 Requirements Worksheet

Now back to the live eSeminar. All CBORD general support e-mail addressees, web address, eSupport phone numbers, individual presenter e-mails, and phone numbers were shared with attendees for any follow-up or clarification. After the session concluded, within 24 hours an e-mail followed with an evaluation form and some bonus items. The session was recorded and available later at CBORD's website.

Additional reference/training materials include a four-page training manual "Handheld Inventory Management System Using HIMS Version 4." Documents are constantly upgraded and supported by online web tools. All training materials and resources are easy to understand as reported by the users and Susan.

Observations of Both HIMS User Accounts . . .

Della reported that they carried approximately 1500 inventory items in their HIMS Palm Tungsten T2s. They provided 55,000 meals a month, approximately 1800 a day. Her operation was a single unit, with a day-care center. They take two major inventories per week and conduct a month end cost. Bread and milk orders are based on par stock, and have an open storeroom (not ingredient based).

Marlene stated that the training on use of the handheld was intuitive. The item was there. All you had to do was count and enter the result into handheld. After you entered the counted items, you synched to the PC, and it communicated immediately, downloading the count and all the calculations. The software calculates your requirements based on menu projections and inventory, creating a Requirements Worksheet (see Illustration 14.4). This grocery list based on par levels already preestablished was a boon to the ordering process. She reiterated that menu- and inventory-driven requirements are consolidated by purchase group. Item name; order quantity; calculated requirements; purchase unit; on-hand, on-order, committed, and par stock levels; and reorder points are all clearly represented on one page. The on-hand, on-order, and committed figures provide a complete status of each item and voilà, the software creates an order. You and/or a production team reviews the order and gives the okay, and the software creates a list for the vendor. This is just one of the valuable forms that FSS creates. Others include production recipes, advanced preparation lists, and pre/postservice comparisons.

Marlene's location did not have wireless and scanning capabilities when interviewed in 2005 but hoped to in the near future. She shared the facts that they "loved" their handhelds and the screens were simple to learn, confirming much of what Della noted. They had instituted FSS in 2002, and in 2003 embraced handhelds in foodservice operations, with the warehouse being the last to sign on. They carry approximately 1500 to 2500 items in their inventory, serve 6000 at one meal, totaling 18,000 meals a day (and I thought 10,000 at URI was a coup!). They have six-plus locations using handhelds, with four retail operations interested in barcoding.

A quick but important observation must be made here. The reason that many have not embraced scanning in the foodservice end-user arena is that there is limited ability to scan products because there are few or no barcodes on items purchased from our foodservice vendors/ purveyors. We are way behind other industries in requiring such scan/barcodes on foodservice-related products that we purchase daily. This is an area where we as users and chefs must advocate implementation so that we do not have to consider adding our own in-house barcodes to ease and expedite this handheld activity. This does not even address radio frequency identification (RFID), an automatic identification method that stores and remotely retrieves data and is the future.

Live eSeminar Geri Menu . . .

In October 2005 an additional eSeminar enlightened us and, although it was not directly related to inventory and ordering, provided a few items of interest since they were handheld related. We were hooked up visually via Microsoft Net Meeting® and verbally by phone conference call.

Verbal and visual cues helped us identify PDA instructions to start programs, Palm desktop solutions, synching Palm software, tools, user additions, hardware options, software recommendations, methods of adding users and system controls, and much more. The agenda covered an intense hour. This type of training in the past would have been costly both in time and travel. It was beneficial not only for the user but also approximately 10 others from varied foodservice locations, who participated and were valuable resources. Everyone had an opportunity to share their experiences and ask questions. The networking capabilities were endless, and this training format is cost-effective, helpful, and timely. It is truly amazing!

A most interesting piece of data shared covered the capabilities of loading 1000 meals for over 500 residents on one PDA Palm-based system (not PC) as the platform. Palms identified included Zire 31, 72, Tungsten E2 (preferred), and T5.

Additional eSeminars may be found at www.cbord.com/online for CBORD users, and these sessions are recorded and available to users to review information shared by all parties, including presenter and audience comments.

D. RESULTS

The physical inventory process of taking a standard inventory on a Palm (handheld) to include generating a tally, printing/selecting a worksheet, taking a physical count, and entering the physical count takes about the same time it normally does when conducting a paper inventory. The greatest savings, though, was realized when accessing, processing, and calculating the inventory.

The user, chef, or purchasing agent may "jump" around on the handheld by storage area, search, and add brand-new items not yet coded. When you sync up again with the ease of a few taps on the screen, it downloads the physical count, calculates the numbers, and downloads your entire FIDF in a brief time, creating an order—a considerable time saver! The training session attendees noted the following "cool" results: those who hate to take inventory now enjoy it, less human-error entries exist, and one can add items on the run.

Della reported that fewer steps resulted in increased accuracy and, of course, less labor. Since handheld implementation in 2001, she has not added labor and has experienced growth, allowing the company to move labor from one department to another. Her FTEs (full-time equivalencies) decreased. When I asked about any improvements to the HIMS product, she commented that there were none and praised its efficiency and convenience.

She was very positive and mentioned that inventory questions/ errors had decreased dramatically. Changes could be made easily in the food item file in Foodservice Suite, and the count by product feature had tremendous ranges to include each, case, count issue, and more. The definition variations were simple and manageable. The math did not have to be performed in the storeroom or refrigerator, and all you had to do was know the item (apples by the case), enter the count (2), and, behold—it calculated itself.

Entering new items was simple due to the FIDF file capabilities already in the handheld download. Many chefs have a love-hate relationship with math. Our minds are on getting the food out safely, on time, and within quality profiles, but we would learn to love it (math) with these tools! I asked Della if they had reduced food costs solely through the use of HIMS, and she said that they had not studied this but anecdotally felt that it might be so.

Marlene shared similar findings and added that forgetting to count items and transposing numbers, specifically, were a thing of the past. She did note that handheld screen size dictated the reordering of the storeroom. The storeroom originally had apples next to pears (and paper inventory sheets did not), but on the handheld apples were followed by apricots, alphabetically, so they changed the order of items in the storeroom to match the handheld inventory. What they realized was saved time and energy for the user. This was something that they were never aware of when they were taking a paper inventory because items that were not sequential, alphabetically or by bin, were hunted for by the person counting rather than sequenced directly. Why didn't they change the storeroom to accommodate the paper inventory sheets? Customs? Tradition? Susan remembers the days when she walked from one end of the storeroom to the other, back and forth and back and forth. The HIMS tool made them acutely aware of this. Brilliant change! The university is working with CBORD for wireless connectivity to remove synching. This will get them closer and closer to a "real-time" scenario. This is very forward thinking and will be even more impressive and productive.

E. COSTS

Rich H. from CBORD was kind enough to provide these costs for a single-user educational grant. The solutions included: Foodservice

Suite (FSS), HIMS software, Non-CBORD software and equipment for one single site, and a single computer or workstation, which totaled just over $10,000 for the first year with licensing fees each additional year. (Note again that this was with an educator's discount.)

Use of non-CBORD systems on additional workstations beyond the initial number of workstations requires a one-time license fee and an annual fee for Sybase program. Use of CBORD systems on additional handheld computers units beyond the initial number requires an additional one-time use fee and an annual fee for the Sybase program.

A breakdown of such charges is confidential and negotiated, since there are multiple considerations that must be taken into account for the personal service and requirements dictated by every end user.

Handheld/personal digital assistant (PDA) hardware is not included. Client services include on-site training, but with excellent web-based tools and eSeminars, this can represent a considerable cost savings in travel, lodging, and out-of-pocket expenses. CBORD has spent the past 30 years increasing its technological solutions with you, the end user, paramount to their mission!

And . . .

This case ends Susan's trilogy of discovery and hopefully starts you, the reader, on your quest to know more about handheld use. Unlike the other two cases, where she actually had face-to-face contact with experts, this case was developed solely using computers, phone, e-mail, an eSeminar and a Webinar (seminar on the web), and faith that a strong technological foundation will help our profession.

Yes, the technology is burgeoning with real-time applications and that is what will make this a perfect solution. What could happen is that when your sous chef says, "we need '_____' (name a foodstuff)," if you have Bluetooth or any wireless capability, you open up your inventory food item file on your handheld, tap on what you need, send it via wireless to your PC, and have it automatically downloaded to your purveyor. It is instantaneous!

A sudden increase/decrease in consumption of a product because of a breaking news story, a march by students protesting veal on your campus, the dreaded or sometimes loved food critic's review, the weather, a food purveyor's mistake or substitution, or a food bioterrorism attack impacts selections on your menu, and with a tap of your stylus you can rectify this.

LESSONS

- It takes a village to raise a handheld user!

- Otters are not solely animals.

- Inventory/ordering modules do exist and are successful, but why are so few willing to talk about them; now I know why.

- Wireless "real-time" inventory/order direct to purveyors must be advocated for foodservice.

- Foodservice is painfully far behind other food-related businesses, such as supermarkets, when it comes to using "just-in-time inventory."

- Barcodes on all purchased food products need to be advocated by foodservice end users.

- eSeminars or Webinars are gems in the rough—with more polishing, they will result in jewels of many colors.

5 QUESTIONS FOR STUDY

1. What does the future hold for inventory/ordering modules for handhelds? Are RFIDs being used?

2. Who might you have to cross-train once you institute handheld inventory systems? What if I am visually challenged or if English is my second language?

3. Where do I go if I drop my handheld and it is broken? What are some additional resources to protect me in the future? Name them and their costs. Are ruggedized handhelds still expensive?

4. When is the best time to institute a handheld inventory/ordering system?

5. Why should I institute handheld computerized inventory/ordering in my foodservice operation? Are there any ethical concerns?

And More Thanks

Appreciation is extended to: Team MICROS (again); George Kotcher, Barrington Software, Cookenpro (again); Traci Hotch, public relations manager, Mobile Computing Division, Symbol Technologies; Dave Feaster, Sales Partner System, Inc; John King, owner, of Sinead

Corcoran, executive chef; Gary Fernbach, network administrator; Jay, Kevin, and Dave at JKING Foodservice Professionals, Inc.; and Gary Leonard, North Shore Health Systems, for furthering my education in the needs and services of additional solution providers, intermediaries, and end users for inventory ordering on handhelds.

George Kotcher has been a constant source of support since Susan's early days at NYIT, when she was teaching a culinary software class. He was her first "guest" lecturer with his recipe management software and more. He entered our web-based synchronistic chat in Blackboard and "talked" with the students on the use of Cookenpro.

In 2005 Barrington, George's company, was testing scanning software with costs in a middle range—not too high, not too low. Users would purchase Cookenpro Suite, a combination of CookenPro Commercial and PDA Kitchen Magician. At this time, PDA Kitchen Magician ran on a desktop synching with a handheld. The barcode scanner (Socket card for example) would run on a Pocket PC–based OS 2000-2003 (not Palm OS) or preferably a Symbol 2800 or 8800 (NOT WINce) that already includes a bar scanner. A Pocket PC could be utilized, such as a Dell Axim ×3 adapted with a Socket scanner card, but some Dells do not include Bluetooth or wireless capabilities. You could, of course, download to your PC with appropriate cables. Palm devices at this time would not support their barcode scanner coding, but the company was investigating implementation, noting that this level of effort would be quite high for Barrington ($$$$$).

As you have been made acutely aware, every technology software provider has standards, and these must be followed so that the integration and communication of all software and hardware can occur for successful completion of data sharing and the processing cycle.

Traci provided us with some Symbol information, and we had so hoped to complete work with them, since they did not reside far from Susan's campus location on Long Island. Dave Feaster was critical in her understanding of the relationships between first-, second-, and third-party users and linking programs, including: interfacing IStocker with a Symbol bundle, NETLink, and matching purchase orders with PDAs (Palm platform). Susan's head was spinning, but she *really* understood why the complete cycle of real-time wireless ordering, deleting, and adding items to inventory was a miraculous event when it worked.

Inventorying/ordering via handheld is becoming more sophisticated and more affordable, but we must keep sharing our knowledge and the lessons learned in online forums that support this growth. Integration with recipe modules and product information reports are on the horizon, so keep pushing for integration. Chefs, advocate for use of these products, test them, and provide feedback (a reminder to all you Microsoft product users that these programs only got better because we, the users, provided feedback). We will we never get to the next permutation, version of software, or hardware ideal that services our specific needs unless we embrace the technology and participate.

We lag behind other professions and their emerging technologies. The foodservice industry and our professional organizations must demand in their mission statements to embrace, learn, practice, and enhance these new technological handhelds and web-based tools. No longer "just-in-time" but "real-time" information is a priority for our industry.

FURTHER INFORMATION

- Barrington Software, Inc.—www.cooken.com
- Dell—www.dell.com
- J. Kings Food Service Professionals, Inc.—www.jkings.com
- Otter Box—www.otterbox.com
- Palm—www.palm.com
- Sales Partner Systems, Inc—www.spsi.com
- Symbol Technologies—www.symbol.com
- The CBORD Group Incorporated—www.cbord.com

ADDITIONAL ACKNOWLEDGMENTS

CBORD Group, Inc.

Roxanne Auble, administrative team leader

Jennifer Christianson, support associate

Tania Bower, support technician II

Heather Hedges, marketing manager—corporate, healthcare, & long-term care

Marty Fischer, senior implementation representative

Lisa Walling, senior sales support specialist

PART **3**

ON BEING A
PROJECT CHAMPION

Chapter **15**

Making Change Happen

NYIT was one of the first culinary schools in the country to introduce the separate placement of potentially allergic food products onto the dessert buffet. This was to protect customers with nut allergies, and Susan led the change.

To do this she had to convince many people at many levels in NYIT's Culinary Arts Center. She discussed the change with back-of-the-house and front-of-the-house personnel and with chef instructors. On Susan's urging, a few read the book *Serving the Allergic Guest* by Joanne Schlosser, which had come out that year. After persuading the curriculum committee, Susan's team introduced the students to the text and techniques.

Good foodservice management requires you to be good at convincing someone to do something. You will need to convince your employees to arrive to work on time, to follow sanitation safety procedures, and, of course, to follow the recipe. You will also need to convince customers to use your services and suppliers to meet your standards. For all of these, you need good persuasion skills. To start with, all you need is love.

All You Need is Love

It is important that you care about the change you are trying to make. People will not care about your argument any more than you do. Therefore, if you are half-hearted in your attempts at persuasion, it is only correct that you not be taken seriously.

Practice Makes Perfect

However, competence is more important than just enthusiasm. And for competence you need practice.

Susan taught multiple and emotional intelligence skills as part of her course at NYIT and has introduced these same skills when modeling a philosophy for the new graduate program at Baltimore International College. The central message is that these skills are not innate, they are learned, they can be improved, and we all need practice.

If you learn best by reading, Susan recommends *Emotional Intelligence: Why It Can Matter More Than IQ* by Daniel Goldman and *Emotional Development and Emotional Intelligence: Educational Implications* edited by Peter Salovey and David J. Sluyter.

For Mohammad, it was Patch Adams' autobiography.[1] Put the Hollywood interpretation out of your mind, and focus on Adams' central skill: how to connect with anyone, instantly. He did this by constantly practicing, and putting himself in unusual situations. He conducted his favorite "assay" in lifts. How many floors would it take for a conversation with a complete stranger in the confines of an elevator to arrive at a personal and meaningful topic? By putting himself in such situations, and trying many different techniques, he developed his own highly effective style.

So, in your career seek out difficult situations as opportunities for practice. When a customer is unhappy with service in your restaurant, make it your business to talk to that person and change the service to their satisfaction. If a member of your staff is trying to change the recipe for a menu item, take advantage of the opportunity to practice persuading him or her to stick to what your team had spent so long perfecting.

Gluten-Free Products

HyVee ©
Employee Owned

to time and Hy-Vee, Inc. makes no representations or warranties whatsoever regarding this list, its accuracy, or the fitness of any product for any particular purpose.

Start

Become a Model Citizen

Getting your first handheld computer is a little like getting your first car—you never really understand what it can do for you until you actually have it. However, where almost everyone expects to own a car as they get older; no similar expectation exists for handhelds. So, how can you possibly explain to someone the need for such a machine when they are so comfortable in their current state?

Start by highlighting functions that no other team member achieves without a handheld. For example, if you cannot make it into work for your shift, impress everyone by your professionalism calling each team member individually to apologize, and then find a replacement. Your Address Book is invaluable for this. It is also great for organizing social events. Made sure that they all knew that this was possible and necessary because you had that machine in the palm of your hand.

Mohammad played on the situation in two ways. First, all his colleagues soon realized that the fastest way to get a phone number was to ask him to look in his handheld. Second, when a new member joined his team he offered to beam them the relevant contact details. If they did not have a handheld computer Mohammad explained to them what a golden opportunity they missed out on. He also promised to beam the details as soon as the colleague bought a device of their own.

Repetition, Repetition, Repetition

Every year, advertising awards are given to the funniest and most innovative adverts. So, why do so few advertisers actually invest in innovative campaigns? Because the industry has known for a very long time that repetition beats innovation every time. A company will sell far more product with a cheap, but heavily repeated, advertisement than it will with an expensive and long one.

So, Susan has started to nag people continuously about using handhelds. Use every possible situation to remind people of how much better their life could be with a handheld computer. Susan likes to think that her reminders are always different, intelligent, and witty. Nevertheless, she fully understands that frequency is their most important attribute.

Understand Every Customer

In a recent survey on the impact of health information, family physicians placed TV as the most influential medium, and saw their own contribution as minimal. When patients were asked, however, they named family physicians as by the far the most trustworthy source, and TV as one of the least. Why do patients value the doctors' opinions so highly? It is partly because the doctor presents the advice in a way that is tailored to that patient's problem.

In the same way, it is important to match your sales pitch to the person you are trying to convince. When Mohammad discussed dietary changes with his patients, he began by listening to them explain their daily routines and why they have those habits. Only after the patient is finished talking does Mohammad feel confident in tackling the habits that need to be changed. He uses words that the patient used and ensures that important aspects of the patient's life are minimally disrupted. For some, the main meal with the family is the most important, so the change is focused on reducing snacks. For others, the ability to regularly chew on something throughout the stressful workday is important. Carrots are great.

Start Low, Go Slow

Susan's vision is for every college and foodservice institution to equip its students and staff with a handheld computer. A foodservice institution that would not do this would appear as silly as one that expects its chefs to pay for their own cutting boards. That is for the future. In the meantime, her aim was to convince just one chef to buy a handheld computer.

After four months with her handheld computer, Susan made such a breakthrough. One of her graduates working in the facility decided to buy a handheld because her son would be joining the culinary program. Susan had demonstrated how useful the device would be for his education.

Four whole months to yield one user for just one machine? It will soon became apparent that this is time well spent. One satisfied convert will convince others far better than you can. In Mohammad's case, whenever a colleague is wavering about making the purchase, Mohammad recommends speaking to his first instructor. Such colleagues always end up investing in a handheld.

Follow the Leader

Many years ago Chef Fred Hendee gave Susan a Chinese cleaver. She showed this to a colleague, who bought one for himself. Soon, employees bought their own Chinese cleavers.

This exemplifies one of the biggest opportunities you will encounter in foodservice when it comes to implementing change—everyone does what everyone else is doing.

We are arriving at the point where the vast majority of foodservice professionals have heard of handhelds, and many know someone who owns one. Showing this book to your colleagues could be the final proof they need that others in the profession consider the technology to be a useful and trustworthy tool of the trade.

The End?

As Mohammad came to the end of his first year of practice, he felt proud of his institution. Many colleagues had bought handhelds, and now departments are making institutional purchases. Most importantly, many of the key educators and decision makers have bought one. A virtuous circle has started, and it will soon become the norm for doctors to have one. Furthermore, nurses and their managers are also asking to buy handhelds, triggering another wave of enthusiasm and changes in working practice.

However, the achievement he is proudest of is that colleagues have started to think differently about what they can achieve at work. Rather than often feeling discontented and impotent about the state of affairs, they see that it is possible to bring about change. They are not waiting for management's next plan to fix things—they have asserted their own power to change.

END NOTES

1. *Gesundheit!: Bringing Good Health to You, the Medical System, and Society Through Physician Service, Complementary Therapies, Humor, and Joy*, Patch Adams, Maureen Mylander, Healing Arts Press, 1998, ISBN 089281781X.

Chapter **16**

Talking to the IT Department

Foodservice professionals in large institutions have some level of access to their IT department. The staff of this department can be a great help as you push your handheld project. However, all too often, foodservice professionals misunderstand and misuse this resource. This chapter explains three principles to avoid misunderstandings, and proposes three practices for dealing with your future allies.

Principles of Understanding

In most large institutions, finding the computer support staff requires a trip to the basement. Not only does this location (accurately) symbolize the importance that the institution gives to these people, but it also creates a distance from the workplace.

1. Pity the People.

Particularly after the dotcom bust, IT staff are underpaid and overworked. In other words, they are just like the chefs, waitstaff, and other foodservice professionals. However, whereas many foodservice staff can complain to each other during shifts and at the bar at the end of the day, computer workers are not listened to.

2. The IT staff should be your friends.

It is important to realize that good foodservice requires good information management, and that in turn requires good computers. The IT department is there to help you in this respect. Like other members of your foodservice staff, the vast majority are hard working and keen to help. If you understand where they are coming from, it is possible to make great use of their knowledge and expertise.

3. But they are still geeks who like cool toys.

It is also important to realize that the computer staff is passionate about technology and all the cool things it can do. "Cool" is important here. Many cool things are useful in a practical way useful, and vice versa. But not always. It is, thus, helpful to be somewhat on your guard when dealing with the "boys" (and it is usually boys) and their toys.

Take the example of picking a handheld computer. Pocket PCs are cool. They tend to have more hardware for surfing the web than Palm-compatibles do. Such capabilities are commendable to be sure but rarely useful on the ground. That's because the surfing is done using either a cell phone or a wireless network card. The former is quite expensive, while many institutions lack the latter. The hardware required for wireless networks is still expensive and difficult to set up. So, buying a machine so powerful that it can surf the web wirelessly is certainly cool but an unnecessary expenditure of money if the network will not reach your institution for another three years.

Practices of Working

So, how do you put these principles into action? Having understood these colleagues, you must also do a little work with them.

1. Understand what you want to achieve before you speak to them.

Mr. Turner, Mohammad's secondary school economics teacher, taught much more than economics. One of his favorite pieces of advice was that you should "Be careful what you ask for. You get what you ask for, not what you want."

In other words, think carefully about what simple problem you would like to realistically solve. The more detailed you are, the more useful the advice will be. For example, if you ask to be able to carry all your inventory records with you, the answer will have to be a computer system that will replace all your paper records. While this is desirable, and many institutions are working on it, the solution will take up large amounts of money and time.

With a little thought, you might decide that what you actually need is an inventory entry form and temperature recording. This need is much easier, cheaper, and faster to address, as shown in the HACCP case study in Chapter 13.

2. Make sure that they are able to support you.

Discussing your desired purchase with your computer support team beforehand will avoid nasty outcomes and permit smooth transitions. An example of a common risk is the compatibility between the handheld computer you want to buy and the desktop computer you already own. All handhelds today require a USB connection, but some old desktops to do not have this socket. Pocket PC handhelds, and those made by Sony, require more effort to work with a Macintosh desktop or laptop. Nor do Sony handhelds work with some versions of Microsoft Windows. The list is rather long and technical but easily digestible by your support staff. Take advantage of this great resource.

More important, even if the new machines will not work with your existing PC, the IT department will often surprise you by how flexible they are willing to be. Their teams are constantly upgrading computers and may happily adjust the scheduling to meet your needs. In other words, just because you currently have an old computer now does not mean that you will always have an old one.

3. Make sure that you are able to support yourself.

Many computer problems can be solved with a little bit of knowledge. Yet many foodservice professionals are too scared of technology to try to acquire this knowledge. This leads to delays as they wait for the IT support staff to come and solve the smallest problem.

This is most amazing given that chefs learn new skills every day by applying the "see one, do one, teach one" principle using sharp objects. Learning on a computer is usually far easier (and safer).

There are many ways to improve these skills, from reading books to taking courses, but perhaps children can teach us the most powerful method: playing. One of the reasons that youngsters are so good at picking up new technology is that they do not find it scary, and instead delight in pressing all the buttons.

Reassure your colleagues that handheld computers are simple to use and difficult to damage. Encourage them to play and learn, and you will be surprised at how many problems you are able to solve by yourselves. Remind them that handhelds are a portable extension of their desktop PC, and if they delete data or programs, it is usually easy to replace these by either beaming or synchronizing.

Chapter 17

Choosing Software for the Team

The great thing about handheld computers is the range of software available. As discussed in Chapter 3, you can have a lot of fun choosing the software with the features you want at the price you can afford.

When it comes to buying software for use by the whole team, however, you must consider a few other aspects. These include simplicity, industry standards, and multiplatform readiness.

Focus on Three Simple Uses

With time, you and your team will find an increasing number of uses for your handhelds. Indeed, as you will read in Chapter 20, this aspect of the machines eases the process of getting funding.

Nevertheless, when you start the project, we urge you to keep things simple. A good rule of thumb is to stick to three simple applications for each project. This rule of threes is not as restricting as you may think. For example, count the organizer functions of the machine as one application. In other words, the Date Book, Address Book, To Do Lists, and Memo are so simple and powerful that we lump them together.

There are several reasons for restricting yourself. First, when it comes to training the team, the learning curve will be gentler. Second, the software itself is more likely to work: complexity makes for crashes. Finally, it is good financial discipline. When making the business case for the purchase, if it is not justified based on simple software that simply works, then you should not try to base it on lots of software that might work.

Stick to Industry Standards

Technology is constantly improving, and advances in hardware and software continue to impress. Usually, your team will upgrade or expand their equipment. But you will soon discover that your most valuable investment is the time you have spent collecting and inputting data.

The only way to protect this is to ensure that you store the data in an industry-standard format. In other words, you must choose a format that the computer industry will continue to support even as new technologies arrive.

To understand this, think about your institution's labor and sales record system. What would happen if today you moved to a new computer system that had none of the records about meals served over the past six months, one year, or seven years? This example is particularly frightening to chefs, who rely on seasonal data for purchasing and menu decisions. These data are perhaps far more valuable than all the money spent on buying the technology.

Do not be fooled into thinking that you will never upgrade. Even if you plan to use the same type of handheld and software, the hardware will eventually break, and you will need to buy new devices. You must ensure the old data travels smoothly to your new setup.

So, how can you tell if your choice today is also the industry's choice tomorrow? Of course, you can never predict this with certainty. Nevertheless, there are a few rules of thumb to guide you. For example, the industry standard software on your PC is likely to become the industry standard on the handheld. In other words, if you are buying word processor software, make sure that it works with Microsoft Word; for accounting software, go for Microsoft Excel compatibility; and for databases, easy integration with Microsoft Access is an absolute must, even if you do not own or use the latter.

Another rule of thumb is to find out what other people are using. Ask your colleagues in other institutions or countries what software they have picked.

Perhaps the most powerful way to address this is to find out the sales figures directly. Handango (www.handango.com), for example, includes a league table of the top-selling software in every category. If your software is in the top three, it is clearly a market leader and, thus, a safe bet. This is particularly useful in cases where there is no PC-industry standard, or if several handheld versions work with the same industry standard. Go with the herd.

Incidentally, this does not mean that you should avoid buying a Pocket PC. It is true that Palm-compatible handhelds began as the industry standard and have more software titles available than Pocket PCs. However, the two platforms are roughly equal among new users, and the power of the machines ensures that foodservice software developers usually make a Pocket PC version. Again, both Palm and Pocket PC are safe bets, and you should just go with the machine you enjoy the most. This brings us to the next point.

Be Ready for Multiplatform

The software you choose must work on both Palm and Pocket PC. For electronic books and some other applications, you can even include compatibility with other handheld platforms.

This is important even if your project begins with buying machines of the same type. This is because in the future it is highly likely that a machine of the other type will be added. Perhaps the machine will be bought by your existing team as you are tempted by new technology. Alternatively, perhaps a new member of the team will have their own handheld computer and would naturally like to use it. Either way, you must prepare for the eventuality.

Even if you never go down this path, knowing that your data is safe brings peace of mind. Look after it, and it will look after you.

Chapter **18**

Training

At a recent conference, Mohammad heard an interesting figure. Apparently, for every dollar spent on buying new computer equipment and software, eight dollars are spent on training the staff. Do not be scared of this. The point is that training should be an important part of your plan. In fact, one of the best reasons for choosing handheld technology is how easy the machines are to use. Nevertheless, as project champion, you must budget time, money, and effort toward training to get the full value out of your team's investment.

The first step is to convince your colleagues that the training is a good use of their time. However, the next step is to make sure that the training actually is a good use of everyone's time. Through our work on handheldsforchefs.com, we have picked up a few pointers on how to make training work. Here is our list of top ten training tips:

1. Brevity.

The first thing a user must learn is how to input data on the machine. This is especially the case with Palm-compatible handhelds, which use a form of handwriting recognition called graffiti. The key aspect of this tutorial though is that it must be short. Spend no more than five minutes going over the technique of handwriting. With Pocket PCs there are four different ways of entering data: the same graffiti as Palm-compatible handhelds, an alphabet that is more similar to normal letter shapes, cursive handwriting recognition, and the onscreen keyboard. The first leads to the fastest handwriting, but most users find it initially off-putting because it requires time to learn. The cursive handwriting recognition option seems the most appealing at first because it promises that no learning is necessary, but for many people the results are disappointing. Train the users in the different alternatives to find the one that suits them best. If you do not, they will revert to the onscreen keyboard, a frustrating though predictable method and a sign that the user has not been trained properly.

2. Activity.

People learn by doing. Encourage the users to practice their data entry with useful tasks. For example, let them write their own address into the handheld.

3. Creativity.

You can have a lot of fun designing tasks. For example, don't just tell the users about the "Find" function. Instead, challenge them to find guidelines on purchasing spirits that are hidden away on your own machine.

4. Relevance.

To ensure that your students have fun, make sure that every task is relevant to their daily work, at least to start with. Often the food-service professional is busy and skeptical, and hence impatient. Only by starting with tasks that highlight the handheld's ability to save time will your student want to spend more time exploring the machine's full potential.

5. Beaming.

At a recent conference that Mohammad spoke at, over 90 percent of the audience already owned and used a handheld. This was not surprising, because it was a computer conference filled with computing professionals. However, as he mingled with them later on, he found out that no one had used beaming before. Not only were they missing out on a huge amount of functionality that they had already paid for, but they were also missing out on a lot of fun. Do not let your students miss out. After they write their first address, ask them to beam it to you. Then beam them the other phone numbers, guidelines, and recipes that you have. They will quickly see the advantages. Mo has taught Susan well, and she passionately believes that beaming equates to sharing and is what humanizes technology.

6. Confidence.

As the foodservice professional who is most experienced with a handheld, and acutely aware of your own team's needs, you are supremely qualified to help train your colleagues. Have faith in your ability to lead the way.

7. Support.

It is always good to have help from your IT department, especially during the first session, because they can help you deal with any machines that crash. As mentioned in the previous chapter, make friends with your IT staff, and involve them in your plans early on.

8. Flexibility.

As shown in the three case studies, there are many ways of gradually phasing in the training. With a little thought, you can find a way that matches your colleagues' schedules.

9. Simplicity.

At the start of your project, pick three simple ways that a handheld computer can help your department. This simplicity ensures that only a small amount of time and effort are necessary for success but greatly increases everyone's confidence that the problems are being solved.

10. Independence.

At the end of your project, make sure that your colleagues know enough to learn on their own. Freedom in experimentation is a construct for learning. Show them how to find, try, and buy software. This, combined with the confidence gained by seeing your three simple applications successfully working, will inspire the members of your team to tackle other problems in the department. *Prepare for your former handheld computer students to teach you new solutions.*

Chapter **19**

Electronic Documents

Each handheld computer can carry an enormous amount of culinary reference information. As mentioned in Chapters 7 and 8, many books are already available to download from the Internet, and your team will find these useful. However, often the most relevant type of reference information is that generated locally. This includes institution's recipes, yield tables, lectures, inventories, and any other information you want to carry with you and share with your colleagues. This chapter begins by showing you how easy it is to make these documents available to your colleagues.

If you enjoy the ease and power of sharing small documents and are feeling adventurous, you might even want to learn about making your own electronic books. With this in mind, the second half of this chapter looks at the history of publishing and discusses how you can publish your own electronic book. After all, every chef has a dream of publishing their "own" recipes or memoirs.

How to Share Documents

There are many ways to share documents with your colleagues, and examples include using Memo, word processors, and reader programs. Each has its own strengths and weaknesses. Often, reader programs are most appropriate because they ensure that the document cannot be changed. For example, whoever is reading a RepliGo document cannot accidentally delete a decimal point from a recipe or budget. Therefore, in this chapter, we shall discuss all three types of document-sharing methods, but focus on reader software.

The Humble Memo

If you have beamed a memo from your handheld to a colleague, then you have already shared a document. Yes, it really is that simple.

Part of the simplicity is because all Palm-compatibles and modern Pocket PCs can accept beamed memos.

With time, however, you will come up against limits. First, the memos are quite limited in length. This is fine for a short recipe but no good for detailed instructions. Furthermore, apart from writing in capitals, you have no other formatting tools. No bold, no large writing, and certainly no tables.

Not everyone finds these limits a problem. Mohammad, for one, still depends on the Memo program for most of his notes. In fact, the discipline imposed by sticking to such short snippets of information is of great benefit to the end user. In other words, reading a concise summary is much better than scrolling through vast amounts of detail on the small screen of a handheld.

The Flexible Word Processor

Usually, this lack of formatting drives users to word processors. The transition is particularly easy because the vast majority of local documents have already been created in Microsoft Word. These files can be easily and quickly imported onto the handheld using Word To Go on Palm-compatibles, and Pocket Word on Pocket PCs.

But be aware of two important issues: First, just because it is easy to import a file, it does not mean it is easy to read the file. What is clearly readable on paper or a large PC screen can involve agonizing scrolling on a handheld. You must try to divide your document into smaller chunks to make it usable on your handheld.

Second, you must ensure that your colleagues also have the word processor program. On some of the older or lower-end Palm-compatibles, this is not the case. Nor is it possible to beam Word To Go files to Pocket Word, or vice-versa. As mentioned in Chapter 17, forcing your team into a single platform is not good practice.

In summary, not everyone like these programs, but you should try them to see how your team gets along.

The Safe Reader Programs

Chapter 6 described two examples of reader software: RepliGo and MobiPocket. The PC versions of both allow you to create read-only documents that look good on any handheld computer. The simple PC versions cost around $20 to $30 but allow the creation of documents on one PC for the whole team's handheld computers.

RepliGo can convert any printable document into RepliGo format. This includes Word and PDF documents but also web pages, Power-Point presentations, and individual recipes from recipe databases. In other words, anything with a "Print . . . " command in its File menu. The conversion is efficient and the results look good on a handheld computer. MobiPocket can convert web pages very well, and to a lesser extent Word and PDF documents. For most teams, RepliGo will be more appropriate.

The more expensive versions—RepliGo PDF Mobilizer for $99 and MobiPocket Creator Publisher Edition for $150 – are appropriate for larger-scale work. RepliGo PDF Mobilizer creates a folder on a PC. If you add any PDF file to that folder, a RepliGo version is created within a few minutes. This is faster than manually opening and selecting Print . . . from its File menu, as you would have to with standard RepliGo. More importantly, if the folder is available to other computers on the network, any of your team members can create the files from their own computer.

MobiPocket Creator Publisher Edition, on the other hand, is great for long documents: safety manuals, guidelines, and that book you have wanted to write all along, for example.

A Brief History of Publishing

In the 1450s, the nations of Europe were undergoing a massive increase in complexity, with developments in science, commerce, law, and warfare. In many cases, the bottleneck for further development was the ability of scribes to produce written documents. No matter how many new entrants to the profession, society still needed more.

Enter Johannes Gutenberg—inventor, goldsmith . . . and businessman. Like any good businessman, Gutenberg was constantly on the lookout for market opportunities. He diagnosed the market's need to mass produce written texts. Also like any good businessman, Gutenberg borrowed money to research a solution for this gap in the market. He produced the printing press and introduced the innovation of movable metal type. Finally, he was good at marketing: his first printed title was the Bible, the world's first, and still greatest, bestseller. However, Gutenberg had an eye on where the real money was—indulgences. These were rich people's way of buying forgiveness from God for their sins, and the Church's way of funding religious wars. Gutenberg wanted to mass produce these indulgences for the Church—in effect, he wanted to establish a licence to print money.

Sadly for Gutenberg, he never got that far in his business model, because he was unable to pay off his creditors in time. His printing press and patents were confiscated. Nevertheless, he had still unleashed a revolution: publishing.

Modern Publishing

Future historians might decide that Harry Potter was the pinnacle of publishing in the twentieth century. The book that became a brand reminded millions of children of the joys of reading and contributed to mass literacy. To Susan, however, the contents of her personal library exemplify the pinnacle.

Susan has books that are obscure. Really obscure.

One book, for example, is *The Anthropologist's Cookbook*, by Jessica Kuper, a birthday gift from her late husband. From her uncle's library, a gift when he deceased, are *The Mushroom Handbook* by Louise C. C., and *The South American Gentleman's Companion* Volumes 1 and 2, by Charles H. Baker, Jr. The latter was published in the year of her birth, and every one of these book tells a story.

Each year, the publishing industry produces thousands more such books, ever larger, ever more specialized, and ever more obscure. Yet every year, the industry manages to make money from such books, allowing further investment. The miracle of modern publishing is that it is still cost-effective to produce obscure texts.

Electronic Publishing

With the advent of electronic books, there is the possibility of further bringing down the costs of publishing. Moreover, the price of handhelds, the platform for electronic book readers, is coming down. Prices are already low enough to allow anyone to publish to everyone.

As a culinary professional with specialist knowledge, you can still push ahead with using the technology. Think about it. You can easily and cheaply produce an electronic book that contains your tips, teachings, recipes, guides, or any other information colleagues may find useful. In addition, the cost of distributing the content is close to zero. Put it on a floppy, e-mail it a colleague, post it on your institution's website, or beam it to your team. The knowledge is available for others to read and use.

Publishing Your Own Electronic Book

There are three steps to publishing your own electronic book:

1. First, you must create the text.
2. Second, you need to convert this text into an electronic book, or e-book, format.
3. Finally, you can distribute the e-book.

The first step is by far the most difficult, but involves the same technology you deal with every day: a word processor. Any word processor will create a .doc or .txt file. To make an e-book though, you will need to convert this into web page language: HTML (Hypertext Markup Language).

You can do this in Word by choosing "Save as Web Page . . . " from the File menu, but the ideal medium to work with is a web page editor such as Macromedia Dreamweaver or Microsoft FrontPage. Web page editors create files in HTML format. Open source software, such as NVU (www.nvu.com), is available free of charge and is an increasingly sophisticated alternative to expensive rivals.

Using HTML is good for two reasons. Superficially, this is because all electronic book converters are HTML-ready, whereas only a few are Microsoft Word-savvy (and none work with other word processing programs). More importantly, the principles of making good links between sections underlie the design of a good electronic book. Making an e-book-friendly web page from a long text document is best done by splitting the document into digestible chunks and creating links between them. Otherwise, your users will need to scroll through long sections.

Having designed the set of web pages that constitute the book, the next step is to convert them into a single document—an electronic book. So far, there are many players in this market, with many competing formats. This is unfortunate, but it does not mean that you have to commit to one format now. As mentioned above, all the software vendors depend on a starting file being in HTML. Furthermore, the conversion process takes minutes (and can be automated), meaning that you can easily publish your document in a form that is suitable for all formats. Simply drag the web pages into the window of the conversion program, and press the publish button.

In 2006, there were four major players in the market to choose from. Adobe (www.adobe.com/acrobat) makes the PDF documents you already know about, but Microsoft (www.microsoft.com/reader), MobiPocket (www.mobipocket.com), and eReader (www.ereader.com) all have their own products. It is not yet clear which the winner will be. EReader has the highest earnings from book sales. On the other hand, all these earnings were less than $100 million in 2006, meaning the market is in its infancy. Adobe's Acrobat Reader has an installed user base of over 150 million worldwide. However, the vast majority of these are on PCs. The Microsoft Reader is the default reader in Pocket PCs and Windows XP, giving it a growing user base. This still means that many potential customers are excluded, because substantially less than half of handheld computers are Pocket PCs.

If there is one take-home message from this chapter, it is that you should not use Microsoft Reader. For all its features, it is not multi-platform, and you should not consider it for any handheld project.

Instead, you might find that the most attractive solution comes from MobiPocket (www.mobipocket.com). A small and innovative French company, it is expanding at a rate that has surprised its competitors. First, it makes its reader software freely available for all handheld com-

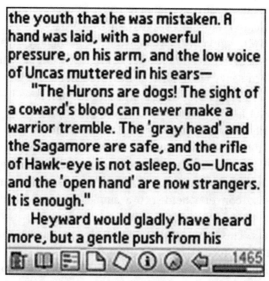

the youth that he was mistaken. A hand was laid, with a powerful pressure, on his arm, and the low voice of Uncas muttered in his ears—

"The Hurons are dogs! The sight of a coward's blood can never make a warrior tremble. The 'gray head' and the Sagamore are safe, and the rifle of Hawk-eye is not asleep. Go—Uncas and the 'open hand' are now strangers. It is enough."

Heyward would gladly have heard more, but a gentle push from his

eReader is a popular electronic book format.

puting platforms, including Blackberry, Franklin, Psion, and Symbian. No other company is doing the same yet. Furthermore, its MobiPocket Creator Home Edition software costs around $30. Upgrading to the Publisher Edition allows faster and more powerful document creation, but this is by no means necessary. For these reasons, Mohammad regularly uses MobiPocket to create electronic versions of medical textbooks.

The final step in the publishing of your electronic book is distribution. If you do not want to encrypt your document, then distribution is easy. Post it on a website, e-mail it to colleagues, or copy it on disks. In most cases, this is the ideal easy and affordable solution.

If you would like to sell the book, or at least control its distribution, then encryption is necessary. Encryption costs money, either in the form of an expensive software suite or licensing fees. Thus, it's usually reserved for commercial texts. In addition, the encryption allows quite a sophisticated degree of control, including restricting the book to a single handheld computer, and blocking all attempts to copy it to another computer. The block is in place even if the same user who paid for the book owns the other computer. This policy has proved controversial, with most users resenting the restrictions. In the United States, such issues are being debated in, where else, the courts.

It will take time to resolve the issues of competing multiple formats and the intellectual copyright controversy. Nevertheless, it is already possible to use the technology to make specialized culinary texts available to all. So, the next time someone discusses making their lecture, paper, or textbook available to students or colleagues—think of using electronic book technology.

Chapter 20

Getting the Funding

If you have read this far, congratulations! We hope that you feel excited and educated about handhelds. All that remains is the small matter of money.

Handhelds are inexpensive, but not free. As discussed previously, it is possible to convince colleagues to make a personal investment in the technology. This is certainly laudable, and most importantly, tax-deductible. Nevertheless, you should not have to pay for such things yourselves, especially software. After all, when you are employed in the culinary and foodservice, are you expected to buy your own computer, small wares, or food supplies?

A key part of your negotiations with your employer is to provide you with the necessary tools for success in your work place and to provide a safe, nutritious, and timely food product to the customer. Although grant money might be complicated for corporate environments, in education there are a few ways of getting small sums of money for innovative projects. One might even be able to create partnerships with education and corporate business, proving with pre- and post-studies the value of handheld use in foodservice. This chapter aims to inspire you to pursue the money to make your own local schemes happen.

Sources of Funding

Apart from your employer and your own pocket, these may be avenues for grants:

- **Achievement awards**: The Institute of Food Technologists may be interested in your technology efforts and have various national awards at webprod.ift.org/IFT/Awards/AchievmentAwards.
- **Food purveyors**: Talk to your local food company representative about funding. They may surprise you by getting involved, espe-

cially if the project can include collecting data for inventory and ordering, and possibly HACCP. In other words, it is a far more satisfying to use your purveyor's money or services. For example, a foodservice team gained funding to equip its account with handhelds as long as the machine included purveyor software that expedited and improved the purchasing cycle. Successful businesses are all about content customers.

▪ **IT companies**: You can often persuade computer manufacturers, software developers, and technology solution providers to supply their products for free or at a reduced price, or negotiate for you with larger companies for a cost savings. The technique is to convince the company that you will help them perfect or market the product for chefs. Remember, those in education, that companies are particularly pleased if they can see positive published papers as the likely outcome. On the other hand, you must be wary that the technology for these trials is usually new and untested. In other words, be prepared to put in extra work making sure things function as you want them to. For example, MTP (My Technology Partners) Inc. has proposed funding a culinary school in a trial to see how culinary students would put the technology to use (www. MTPonline.net). CBORD, Inc. also is willing to work with hospitality and culinary schools with software and hardware costs.

▪ **Professional organizations**: Check with your local or national chapter. These may include the American Culinary Federation, International Culinary Association of Professionals, International Council of Hotels, Restaurants and Institutional Education, Women's Foodservice Forum, and Women Chefs and Restaurateurs. Professional organizations promote their professions by developing future professionals.

▪ **Charitable foundations or assistive technologies**: The United States has a highly active charity sector, raising money nationally and locally. This may be a more difficult area to tap, yet a business case for using handhelds to improve the management of nutritional, food safety and security data is valuable research for the world at large. The NEC Foundation of America (www.necfoundation.org) is a source for assistive technologies for people with disabilities.

▪ **Educational grants**: Culinary institutions, colleges, and universities may consider you (especially if you are alumnae) in your project if school and/or business benefactors and foundations believe that it contributes to the education that they provide. Sometimes we forget that high schools are quite active with educational grants, and partnering with them might prove quite rewarding. However, make sure that your project integrates closely into their syllabus and is clearly auditable. For example, part of the money for certain trials may come from educational grants because the machines will be used in teaching. Palm has Palm Pioneer Grants at palmgrants. sri.com/grants.html with additional support for educational at www.palm.com/us/education. Integrating Handheld Technology in the Classroom grants may be found at handsonlearning.aiu3.net. HP has various technology grants (www.hp.com/hpinfo/grants/us/ prgrams/tech_teaching/index.html) as does k12handhelds.com/grants. php. A new kid on the block, Edutopia (www.edutopia.org/foundation/ grant.php), is a personal favorite of Susan.

- **Employer**: Never hurts to ask. Educate yourself, and then present the value of handheld use after reading this book. They may have funding for tuition, continuing education, or other awards for innovation. And, of course, contact us. Opportunities for and examples of innovative uses in government[1] and health-care[2] facilities may exist with Palm. Foodservice operators participate in all these arenas and this networking may prove fruitful. Innovation funds—the increasing IT budgets available to government and health care has produced increasing innovation funds. The key word here is innovation, which greatly favors handheld computers, as long as you can show a clear contribution to health care as seen in the employer bullet above. For example, a foodservice department may apply for a grant to use handhelds, ensuring that dieticians in foodservice receive accurate diet information from nurses when patients are checked into the hospital. This will help overcome the language difficulties of a new intake.

- **Local community groups, churches, and scout organizations**: Culinary, and food is very popular, and you may want to introduce a handheld badge for scouts or help your church group with healthy managed meals for church suppers and more. No idea is too crazy. It is the community that you support that may support you.

- **Research grants**: Even if you cannot convince the IT companies to pay for your trials, consider applying for a research grant. For this, you need some clear measurable outcomes and, ideally, comparisons between pre- and post-study results on effective use of menu, inventory and ordering, and HACCP tracking on handhelds and the effect on efficiency and cost savings in foodservice operations. There is growing interest in the rigorous assessment of the technology's efficacy in these areas in multiple accounts and affordability for smaller entrepreneurial locations. Newer applications with bedside menu entry, nutritional awareness, and radiofrequency identification (RFID) in tracking the food supply are all potential opportunities.

You may want to place your name on a grants alert website (www.grantsalert.com) or subscribe to a technology grants newsletter.[3]

Find Out What the Funders Want to Fund

During Susan's last few months in the writing of this book, she had the pleasure of meeting Thomas A Capone, Chairman and CEO of MTP (mtponline.net). MTP originally started as Maritime Technology Partners growing with new additions, My Technology Partners, Inc. and My Teaching Partners (www.myteachingpartners.com). Tom invited Susan to work with them as a Mobile Virtual Network Provider (MVNO), modeling Disney, ESPN, and EarthLink, but with a focus on food, culinary, and hospitality. Collectively, by using handhelds and other technologies, service learning will be personalized for each learner with live sessions, simulations, self-directed elearning, videos, and video streaming 24/7. This opportunity will be an eye-opening roller coaster ride as we push to launch My Tasting Partners. Contact Susan at Susan@MTP-USA.com for further details.

From Mohammad's expertise and book *Handheld Computers for Doctors*, one quickly learns that every participant's wants from a project must be met. Read that last sentence carefully, because every word contributes to the success of any project. In other words, managing many participants involves constant assessment, not solely knowing what people wanted, rather learning what they want. Many phone calls, e-mails, and personal meetings are required when asking chefs, vendors, educators, and managers what it will take for them to participate in funded projects. You, as the project coordinator, must think of how to carry out all their individual requests. Getting them to agree to join in takes special skills.

If you do not have the time for training, you may adapt a slower pace to gain funding for projects. In fact, funding agencies often make things even simpler because their application forms describe exactly what the funding is designed to support. Take advantage of this information.

First, you can ensure that you are not wasting your time applying for money that cannot be given to you. For example, some grants are only available to dietitians, so a chef's participation might not be advantageous. More powerfully, you can change your proposal to allow success. You could expand your project and collaborate with a dietitian. Everybody wins. Check out the American Dietetics Association website on occasion for possible partnerships. For example, search for "technology" on www.eatright.org.

There Are Many Uses for Your Handheld

A key part of tailoring your funding proposal is to realize that your handheld has many functions. It is easy to come up with a way to satisfy the grant criteria. Note, however, that this does not mean you can only use the machine for the function for which it was funded. For example, take the dietetic team mentioned earlier. They will want the machine for a variety of reasons, including sharing accurate patient food dietary needs, food sanitation and safety, and reading various texts. To qualify for the foodservice or IT company funding, they might add the task of inventory and ordering specific foods for required diets (allergy-free or low sodium or . . .). This task incorporates the chef, so receiving the funding opens up the populations that may benefit. Of course, the machines would be used for diet audits, but they would also take full advantage of all the other features. Everyone would be happy with this arrangement.

You can also push this principle to qualify for multiple sources of funding. The team could also apply for a small educational grant if the machines are used to teach culinary students about low-fat or low-sodium recipes. They might qualify for an innovation grant if the auditing project produces a database that was freely available for reuse by other dietitian and chef teams around the country. Moreover, the audit could be expanded into a clinical trial for dietitians and, thus, attracted other research funding. With a little imagination, it is possible to get more funding than you originally needed.

How to Sell "Folgers"

If you have identified a need for the handhelds and a use that would qualify for a grant, now you must sit down to write the proposal. Understandably, many chefs shy away from this task and may not even know of such opportunities unless they have been in the academic or nonprofit sector. This is partly because of the time and effort required within an already stressful workday. However, perhaps the biggest barrier is not knowing how to write such a proposal.

There are many ways of circumventing this. First of all, talk to your colleagues who have received funding, or better still, those who work on grant allocation committees. They are usually happy to help and a little advice goes a long way. Second, you can read a book on applying for funding and many are available, including *Grant Writing for Dummies*.[4] Third, there are many excellent seminars and training courses, including those from websites like fdncenter.org/newyork/ny_training.html and www.tgci. com. Make sure that you visit your local library or state resources. The greater your skills, the easier it is to close the deal quickly. A sales background may be quite beneficial in your quest for funding.

In the meantime, you might like to take a leaf out of Proctor & Gamble's book on marketing.

Proctor & Gamble

It is highly likely that you are a customer of Proctor & Gamble (P&G). Looking around your house should produce several items that are sold by the company. Folgers and Millstone coffee, Pampers, Head & Shoulders shampoo, MaxFactor makeup, Hugo fragrance, Sunny Delight juice, and Pringles chips are just some of the brands that it owns. P&G is worth over $100 billion.

A large part of their success lies in their marketing skills. Over the decades, they have developed a formula that has contributed to all their advertising efforts and ensures such consistently high sales.

This formula can help you, because a grant application form is, after all, a sales document. You are trying to convince someone somewhere to give you some money. But don't worry: this does not mean that you should insert photos of scantily clad individuals into your proposal, or include one-liner jokes. Nevertheless, the formula's four principles—drama, promise, demonstration, and result—can still provide a lot of guidance.

Drama

In coffee advertisements, the drama is that waking up in the morning is a difficult task and that coffee may not be good for you. Coffee products may suffer from low sales because inadequate information and support perceives coffee as a "drug."

The drama of your application form is the chefs need section—we need coffee because. . . . The aim is to convince the reader that there is a serious and urgent problem to eradicate this perception about coffee, and in P and G's case Folgers.

Do not hold back emotionally when describing this problem. Mention all the consequences of not having the product and equipment that you request for high-quality production of this beverage. Include the impact on customers, relatives, staff, colleagues, and anyone else affected, especially chefs!

As a chef/manager, you should also ensure that you have the numbers to support your emotive case. For example, including a foodservice incident form (coffee sales are down) is a good way to support your anecdotes. Results of local and national studies take a little work to assemble but are well worth it. Finally, do not forget to enlist the help of your community, school, or college librarian. As well as helping you search the food and beverage service, hospitality, and nutrition literature, they can provide you with local, national, and regional demographic data that highlight your needs.

In most cases, it is worth adding legal and financial ammunition to your cause. This works particularly well when discussing with your managers the impact of the problem on meeting the requirements of your employer, clients, and customers. This is what they are paid to worry about.

Finally, it is worth mentioning that you must make the case clear to the reader. Do not assume that the person assessing your application will be a chef, manager, purveyor, or even working in the same field as you. Assume intelligent ignorance, and you will avoid irritating your audience.

Promise

There is no point discussing a problem that is insoluble, or takes too long to attack. Thus, having scared and depressed your audience with the drama, you must now lift their spirits by promising a way out.

The Folgers website promises various products, roasts, and services. Services include surveys, sharing your thoughts, contact information, product history, wake-up clubs, brewing techniques, recipes, and product locators. This is an excellent example of a goal for a product, or in your case, a project. A version could be to ensure the greatest beverage service in a region with proper ordering, inventory, safety, and menu control.

The corollary of this principle is that you should not mention any woes in your drama that you cannot overcome with your promise.

Demonstration

Folgers is so successful because of the great advances in coffee production and preparation techniques. The Folgers website features services that are tailored to users' specific needs and individual preferences.

Your own demonstration should, thus, include a clear explanation of how you plan to solve the problem. Although television advertisers have very little time to explain themselves, Folgers web advertisement on the other hand is complete, and as a chef seeking funding, you are encouraged to spell out your solution in detail. Not only does this make life easier for the grant assessor, but it also increases the chances of your project eventually working.

The latter is true because the person reading your application assumes that the more concrete your details, the more thought that was involved in coming up with this plan. The more thought you have invested, the more likely you are to be serious about the solution. You are not just asking for money indiscriminately; you are requesting fair support to solve an important problem.

You too will benefit from this intellectual investment. By thinking carefully, and for long periods about your project, you will move quickly and effectively when the money finally appears.

What does this mean in practice? Do not just estimate that six hand-held computers costing approximately $400 will do the job. Rather, pick the machine you like the best, print off its specifications and price, and include them in your application form. Incidentally, with the constant drop in prices, the money that eventually arrives may be more than the actual cost of the machines. Feel free to use the extra money at your own discretion, for example, for buying additional software.

Furthermore, use information from discussions you have had with your team and with other departments that will be involved in the project. Include a feasibility study from your IT department. Highlight positive outcomes from other teams around the country and abroad who have already blazed the trail.

Grant success is in the details.

Result

Results are about objectives. Folgers might claim to be "the best part of wakin' up™," and your proposal might claim a 50% reduction in inventory, ordering, and menu entry errors.

It is important not to confuse objectives with goals. The latter is a desirable statement, while the former is a measurable outcome. Grants allow the pursuit of laudable goals but are allocated based on measurable objectives. The money does not go to those who feel underresourced but to those who can prove good spending.

In some cases, this may be the hardest part of the application form. This is because some things are difficult to measure. In the example of the foodservice departments, improved customer perceptions may be difficult to gauge, but a reduced cost or increased sale for each transaction is more easily reproducible and measurable. So, with a little creativity, you can always come up with a measurement, even if it is indirect.

Finally, long after you have received the money and carried out the project, do not forget to follow up with the results. Contact the provider of the funds and let them know of your success. Some providers will be expecting to hear of the results through the publication of a paper or report. However, even if they do not explicitly ask to be kept informed, a little effort to maintain contact will go a long way, and the person who allocated the funds will feel good about the decision.

Besides, they might even suggest how you can receive more funding for your next project.

The Beginning of the End or the End of the Beginning?

Congratulations! You have reached the end of the book, and hopefully the beginning of your project. We wish you great success but also offer help. If you need any advice or encouragement, you are most welcome to contact us. And, when you succeed, it will bring us great pleasure to hear about it.

In the meantime, take care, good luck, eat well, and share with others.

—Authors@handheldsforchefs.com

Accompanying Website

Handheldsforchefs.com provides further information to accompany this book: to chat, inform, share, or learn more, visit www.handheldsforchefs.com.

END NOTES

1. www.palm.com/us/solutions/government
2. www.palm.com/us/solutions/healthcare
3. www.technologygrantnews.com/grant-index-by-type/educational-technology-grants.html
4. *Grant Writing for Dummies*, Ben Browning, ISBN 0764553070.

Index